Routledge Revivals

The Rendering of Geographical Names

First published in 1957, *The Rendering of Geographical Names* is a clearly presented survey of the little-known borderland between linguistics and geography. Works on toponomy, that is, on the study of place-names in particular countries, are numerous; but works on geographical names, that is, on the names of places and of geographical features used in foreign languages, are rare. But if foreign news is to be reported in English newspapers or discussed in English books; if foreign books are to be catalogued in English libraries; or if foreign countries are to be shown on British maps and in British atlases, the need for rational ways of rendering the names of parts of the world and of places abroad is immediate. And many such names are written abroad in unfamiliar alphabets and other scripts. This is an important historical reference book for students or geography.

The Rendering of Geographical Names

M. Aurousseau

Routledge
Taylor & Francis Group

First published in Great Britain 1957
by Hutchinson Library

This edition first published in 2025 by Routledge
4 Park Square, Milton Park, Abingdon, Oxon, OX14 4RN

and by Routledge
605 Third Avenue, New York, NY 10158

Routledge is an imprint of the Taylor & Francis Group, an informa business

Publisher's Note
The publisher has gone to great lengths to ensure the quality of this reprint but points out that some imperfections in the original copies may be apparent.

Disclaimer
The publisher has made every effort to trace copyright holders and welcomes correspondence from those they have been unable to contact.

A Library of Congress record exists under LCCN: a 58008623

ISBN: 978-1-041-14676-6 (hbk)
ISBN: 978-1-003-67561-7 (ebk)
ISBN: 978-1-041-14679-7 (pbk)

Book DOI 10.4324/9781003675617

THE RENDERING OF GEOGRAPHICAL NAMES

by

M. AUROUSSEAU

M.C., B.SC., F.R.G.S.

SECRETARY, P.C.G.N. 1936–1955

HUTCHINSON UNIVERSITY LIBRARY

LONDON

HUTCHINSON UNIVERSITY LIBRARY
178–202 Great Portland Street, London, W. 1

London Melbourne Sydney
Auckland Bombay Toronto
Johannesburg New York

★

First published 1957

Stephen Austin & SONS LTD
ORIENTAL AND GENERAL PRINTERS
HERTFORD · HERTS

In continuation of the work of

Bradock Mead, *alias* John Green *	*c.* 1685–1757
Rear-Admiral John Washington	1800–1864
The Reverend Isaac Taylor	1829–1901
Alexander Knox	1849–1912
Professor George Goudie Chisholm	1851–1930
Major-General Lord Edward Gleichen	1863–1937
Arthur Robert Hinks	1873–1945
John Hardwick Reynolds	1877–1937

* *See* Crone, G. R. and R. A. Skelton, English collections of voyages and travels, 1625–1846, *in* Richard Hakluyt and his Successors, *Hakluyt Soc.*, ser. ii, 93, London, 1946, pp. 98–109; and Crone, G. R., *Imago Mundi*, 6 (1949), pp. 85–91 and 8 (1951), pp. 69–70.

CONTENTS

LIST OF MAPS

CHAPTER I

THE STUDY OF NAMES

A **name** is a sound or succession of sounds in speech, used to identify a particular thing. Names which consist of single but not necessarily simple words are known as nouns in grammar. They are classified as common nouns if they apply to kinds of things, and as proper nouns (or **proper names**) if they apply only to particular things of a kind. My pet animal, for instance, is of the kind called 'cat', and my own cat I call 'Bill'. Bill is a proper name. It belongs to Bill, and, in his small world, no other cat will answer to it. Dover is a geographical proper name. But 'Straits of Dover' is not, correctly speaking, a name. It is a geographical expression containing a proper name. Slight variations in the sound or spelling of a name may come into use in certain circumstances, and, if they be not wide enough to prevent the recognition of the thing to which the name belongs, may be regarded as **forms** of the name. Roma (Latin) has become Rom in German, Rome in French and English, and Rūm in Arabic.

The study of names as sounds belongs to the science of phonetics, which investigates all speech-sounds. The scientific study of names as names, that is, of the human habit of naming things, is the science of onomastics (Greek ονομαστικὸς (onomastikos), 'pertaining to names'). It represents a very ancient interest, particularly with regard to the names of persons and communities. Freud said: 'For primitive men, as for savages to-day and even for our children, a name is not indifferent and conventional as it seems to us, but is something important and essential. A man's name is one of the main constituents of his person and perhaps a part of his psyche'.[1] And this, perhaps, is why persons and communities resent and resist interference with their names. Christian baptism alone is enough to demonstrate the importance of personal names.

The importance of names has been understood in Chinese philosophy ever since the time of Confucius, for he it was who first formulated the doctrine of the rectification of the names of persons and things.[2] What he meant was interpreted by Hsün Tzu as 'the correct use of terminology'; in the first place, a correct ethical valuation, which Hsün Tzu carried so far that he finally reduced 'the whole problem to the moral task of self-control. Only the man who has his passions and desires well under control, and thus is able to conform himself to Tao, knows how to rectify his terminology, or, in other words, to have the true logic and the right ethical judgment'.[3]

In spite of their antiquity and the part they play in life, names in general, excepting place-names, have been studied with scientific method only in recent years. The subject of Onomastics is not yet adequately treated in encyclopædias, but attention is directed to the articles on 'Names, Proper' in *The Penny Cyclopædia*, xvi, London, 1840, on 'Name' in the *Encyclopædia Britannica*, 11th and 14th editions, on 'Place-names', *Ib.*, 14th ed., on 'Name' and 'Place-names' in *Chambers's Encyclopædia*, new ed., 1950; and on 'Namen' and 'Onomastik' in *Der Grosse Brockhaus*.

The recency of general interest in names is reflected in the change of title of the *Zeitschrift für Ortsnamenforschung* ('journal for place-name study'), founded in 1925, to *Zeitschrift für Namenforschung* ('journal for name-study')

[1] The Basic Writings of Sigmund Freud. Translated and edited by Dr. A. A. Brill. *New York*, 1938, p. 893. (*Totem and Taboo*, 1919, Chapter 4).

[2] Confucius, Lun-yü, xiii, 3.

[3] Duyvendak, J. J. L. Hsün-Tzu on the rectification of the names *T'oung Pao*, sér. 2, xxiii (1924), pp. 221–254.

in 1937, and in the assembly of an international congress of persons interested in names of all kinds at Paris in 1938. A 'Centre International d'Onomastique', established at Louvain, was founded in 1949 on the recommendation of the Third International Congress of Toponymy and Anthroponymy (Brussels, 1949), and is affiliated indirectly to UNESCO. The American Name Society was founded at Boston in 1952 and *The New Century Cyclopedia of Names* (3 vols., New York) was published in 1954. Contemporary interest in onomastics is absorbed mainly in the investigation of place-names (**toponyms**) and personal names (**anthroponyms**), but the Centre International d'Onomastique exists for the study of names of all kinds.

The study of place-names (**toponymy**) has for long been a developed subject with a rigorous technique. It is essentially the historical and documentary investigation of the names used in the language of a particular country for the inhabited and formerly inhabited places, uninhabited places (**lieux-dits**) and geographical features of that country. Ireland, Norway, Sweden, Denmark, Holland and Belgium now have official organisations for the study of their place-names, but the English Place-Name Society, which does comparable work, is an unofficial body.

The names investigated by toponymists are, strictly speaking, geographical names; but the interest now embraced by the expression 'geographical names' is both wider and shallower than that of toponymy. English geographical names consist not only of the place-names of England, but also of the names and forms of names established in the English language for parts of the world, geographical features and places outside England. In short, place-names are the names of places in particular countries, and geographical names are the names of places in particular languages.

CHAPTER II

GEOGRAPHICAL NAMES

A **geographical name** is the proper name, or geographical expression, by which a particular geographical entity is known. Anything which is interesting on account of its location, position or distribution on the face of the Earth is appropriate for geographical study; and any part of the world, feature of the Earth's surface, organic aggregate (coral reef, forest, tribe, sect, race, communion, etc.) or bounded unit of organisation (independent country, part of such a country, diocese, city, borough, farm, etc.) is a geographical entity. The limits of the Earth (its poles, circumference, hemispheres and zones); the parts of the world (its oceans and seas, parts of seas, its continents, countries and parts of countries); its principal lakes, rivers, mountains and passes; its peoples and languages, its capital cities, great sea-ports and grander works of man, have names in the languages of all civilized nations.

Isaac Taylor's *Words and Places* (first ed., London, 1864) and J. W. Nagl's *Geographische Namenkunde* (Vienna and Leipzig, 1903) are amongst the very few works which deal generally with geographical names of all kinds. The emphasis in both is on etymology rather than geography. The only organizations which deal specifically and exclusively with geographical names are the United States Board on Geographic Names (USBGN, 1890) and the Permanent Committee on Geographical Names for British Official Use (PCGN, 1919), but the Economic and Social Council of the United Nations (ECOSOC) has them under consideration. Certain organizations, some temporary, others permanent, in Australia, Canada, Indonesia, Kenya,

New Zealand, the Philippines, Rhodesia, Scotland and South Africa, etc., set up as geographical boards or committees, deal with the propriety and priority of place-names in those countries, and with their official spelling, rather than with their toponymic investigation, and do not concern themselves with names abroad.

The **names of countries** are commonly subject to confusion with the **titles of the states** to which the countries belong. The United Kingdom of Great Britain and Northern Ireland is not a country or a geographical entity of any kind. It is a political entity, a union of peoples living in the countries of Great Britain and Northern Ireland, and the so-called name of the union is really a title. Such official expressions as 'Canton et République de Genève' and 'Commonwealth of Massachusetts', are not geographical names but titles of political entities. Title, unfortunately, is a difficult word. We talk of 'titled people', but the word 'lord' represents rank, not title. In an expression such as 'Lord Derby', Derby is that to which his lordship is entitled. He is the Earl of Derby; and the 'of' in political expressions which are titles rather than names, often reveals them. The polity is designated, and the territory to which it has title is named, as in 'Borough of Hampstead'. Hampstead is a place, now a delimited area, a little country; and the borough is the incorporated ratepayers who maintain it in habitable condition.

It is often important to distinguish between the polity and the space which it occupies. The United Kingdom trades with the Kingdom of the Netherlands, but goods are shipped between Great Britain and Holland. It is common, nevertheless, to allude to a people by personifying the name of their country. We may say 'Italy', meaning the sovereign Italian people. When the French say 'la France' they are not alluding to the country but to themselves, by personifying the French Republic. It is therefore unfortunate that several states have tried in the recent past to induce us to renounce our established English names for their countries in favour of what they consider ought to be the English names. The instance of Siam will be examined below. Persia made representations, before the war, in favour of the name Iran; and there were efforts in Holland to persuade us to say 'the Netherlands', and 'the Netherlanders' instead of 'Holland' and 'the Dutch'. The name 'the Netherlands' applies historically, in English, to all the Low Countries, from Holland to the country about Dunkirk. Abyssinia provides another instance, for there are some who would insist that we call it Ethiopia. But Ethiopia is the name of a country known to the ancient world, and it did not lie within the same limits as modern Abyssinia. Dr. Johnson called his Rasselas 'prince of Abissinia', which surely established the name as that of a country in English. The state to which the country now belongs should properly be styled in English 'the Empire of Ityopya', if we are to render its Amharic name; and this would ensure the distinction of the English from the foreign official name of the country, and of both from that of ancient Ethiopia.

The relationship between the title of a polity and the name of its territory is not necessarily clear and simple, as the following quotation on tribe (*fis*) and tribal territory (*bajrak*) in Albania may show. The *bajrak*, according to Steinmetz, is not a subdivision of the clan or tribe. The two words *bajrak* and *fis* 'stand at the most in fortuitous relation to each other. The tribe is a complex of families which trace their descent from the same ancestor; the foundation of the tribe is therefore kinship, no matter how remote, and this excludes intermarriage. The *bajrak* on the other hand is a community of families living in the same area; the word applies therefore to territory. This general definition, however, needs occasional modification'.

Confusion between the name of the polity and that of the area within which it exists probably extends, at least in European countries, right down to the smallest units of administration. London, for instance, is a name which we give geographically to the continuously paved and built-up area occupied by an organized assemblage of people established at the head of navigation (for sea-going vessels) of the Thames. But the name belongs, historically, to the small area administered by the Lord Mayor and Corporation of London, which we call the City of London. Yet we give it also to an area which includes both the city and the metropolis, an area administered by the London County Council, and called in full 'the County of London'. I doubt that 'city', in City of London, only means the citizens, or that 'county', in County of London, only means the electorate which returns the London County Council. Every small cluster of houses in the country which has a name is likely to be found to lie in a defined area (parish) of the same name. It matters little in daily life and talk, but may mean much at law; and when the issue is between the title of a state and the name of the country in which it exists, we should at least endeavour to be clear.

Many so-called names even of geographical entities are not really names at all. 'King George VI Sound' and 'Mount Misery' are **dedications.** The commemoration of deities, saints, monarchs, events and circumstances by geographical dedication is a common observance. 'Bay of Naples' is a mere **designation** of a thing by reference to something else. 'Long Island' is a mere **description.** It is usually expedient, nevertheless, to treat geographical dedications, designations and descriptions as if they were true names, and I shall do so henceforth.

A geographical entity may have more than one name in the language of the country in which it is situated

(Thessaloniki, official; Saloniki, popular), and may even have a nickname (the Hub, for Boston). In a country where more than one language is spoken it may have two or more official names (Bruxelles, Brussel). A geographical entity which extends into two or more countries may have official names in each: Liqen i Prespës, Albania; Λίμνη Μεγάλη Πρέσπα (Limni Megali Prespa) Greece; Преспанско Језеро (Prespansko Jezero), Yugoslavia.

It must never be forgotten that a name is primarily a sound or sequence of sounds, and will therefore, in civilized countries, have both **spoken** and **written forms.** The written form is a mere record of the name, but it may become a fixed visual symbol carrying distinctions not heard in the sound. Basseterre and Basse-Terre, Georgetown and George Town, St. Johns and St. John's, are respectively the names of different places. There is room for such differences as Bethanie (Transvaal) and Bethany (Orange Free State). The name, being a speech-sound, is subject to change with time (Wymondham, now pronounced Windham). It is also liable to official replacement by another name or succession of names (Oslo, formerly Kristiania; Leningrad, formerly Petrograd, earlier Sanktpeterburg).

Their grammar and elements. A geographical name may be **simple** (York) or **compound** (Isle of Wight). A compound name may have a **generic part,** that is, a part which tells what class of thing a geographical feature is; and a **specific part,** that is, a part which tells which thing, of the indicated class, is meant. The generic part is usually a common noun (Cape), but the specific part may be a proper noun, an adjective, or even a phrase (of Good Hope).

Some names exist only in the **definite state** (*the* Naze); others may be used in the definite state in speech or in passages of writing (*the* Atlantic Ocean), but in the **indefinite state** when written in isolation on charts or maps (Atlantic Ocean, Alps); still others may exist only in

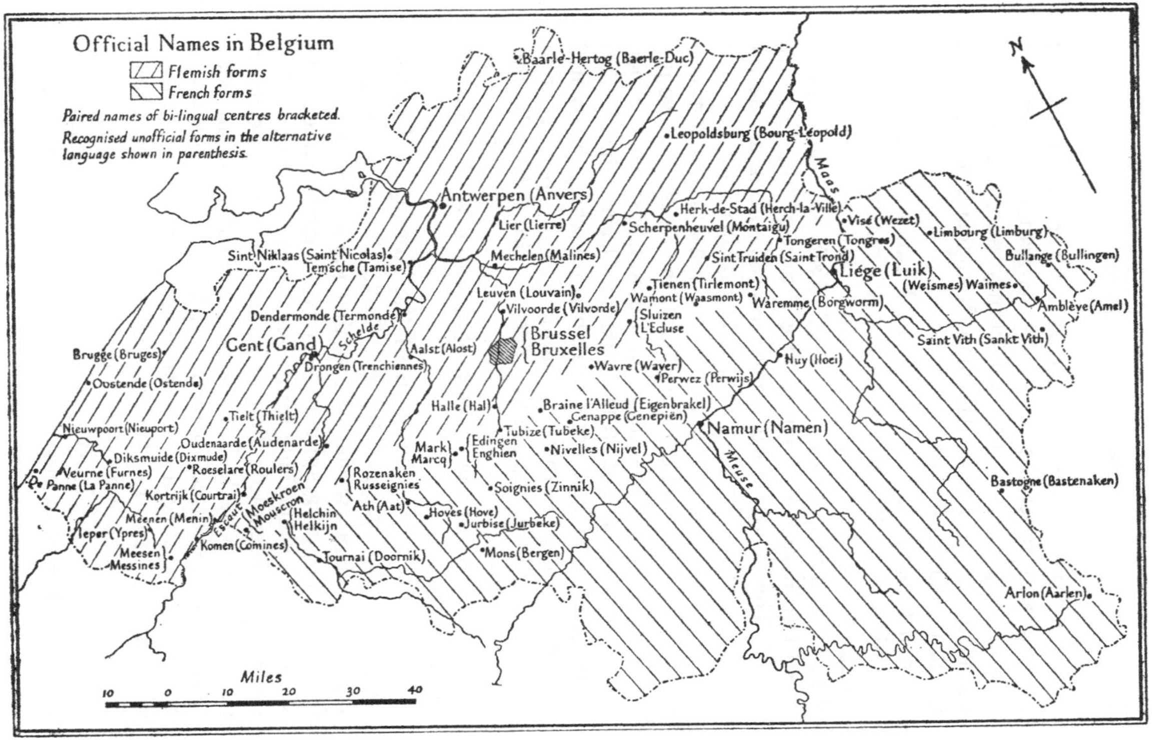

FIG. 1.—*Official names in Belgium.*

the indefinite state (Dungeness). The existence of definite and indefinite states of geographical names in certain languages is not yet clearly appreciated. It is unusually important in Albanian, Romanian, Somali, and the Scandinavian languages.

English usage is not, so far as I can see, systematic. We talk about 'the Straits of Dover', but not about 'the Spithead'; we say 'the River Ouse' but not 'the Cape Wrath'; and in London we talk of 'Walbrook' but also of 'the Fleet River'; of 'Oxford Street' and of 'the Edgware Road'. The nearest I can come towards stating a rule for English is to say that, (*a*) we use the article in the names of geographical features when the name consists of a common noun only, and is unique (the Naze, the Skaw); (*b*) we use it when we drop the term from a compound name which leaves it still unique (the Atlantic *Ocean*, the Sahara *Desert*); and, though not invariably, (*c*) when the 'name' is a geographical expression containing no proper noun ('the North Foreland', 'the Cape of Good Hope'). But there are numerous exceptions to any rule (North Cape; and 'Walbrook', above, was probably 'the Wall Brook' in earlier times).

The Minister of Education of the Norwegian Government in Exile informed the P.C.G.N., during the war of 1939–45, that Norwegians know that the names of certain features in the geography of Norway are correct in definite form, others in indefinite form; but they do not know why. I have also been told that geographical names occur in both states in Greek, but that good speakers of modern Greek, though correct in their usage, are not conscious of making the distinction. In Albanian there are occasions when it is right to say 'Tiranë' (indefinite nominative) but others when one must say 'Tirana' (definite nominative).

Names in some languages have **gender** (*der* Rhein but *die* Elbe; *le* Rhône but *la* Seine), and others may also be

inflected (*nad* Bug*iem*, for *on the* Bug). An Irishman does not go back to Eire (nominative case); he goes back to Erin (dative case). It is noteworthy that the proper names of mountain-systems and mountain-ranges often appear in the **plural** in European languages. This happens in English when we drop the term 'mountains' from the name (the Rocky Mountains are now called 'the Rockies', and the Pyrenean Mountains are now 'the Pyrenees'). The Beskids (ranges on the northern face of the Carpathians) are 'die Beskiden' in German and 'Beskidy' in Polish, both plural forms.

Although the **derivation** of many geographical names (London and Edinburgh, for instance) is obscure, that of a very great number can be reliably traced, and most names prove to come from linguistic elements which once had important meaning. The meaning of Ox/ford, Cam/bridge, Ports/mouth, and innumerable other names, is immediately clear. But that of a great many, like Bea/chy, is not. It is now accepted that Beachy is our corruption of *beau chef*, (imposing head), the expression used by French sailors to identify the feature. These examples suffice to show that many English names which are now written as simple proper nouns consist really of the two parts, generic and specific.

The important **elements** of English place-names have been collected by the English Place-Name Society in a list which is a key to the interpretation of the map of England, though, as most toponymists would add, it is to be used with caution by amateurs. The meanings of many old names have lost all geographical force, either because they are no longer understood, or because local conditions have so changed as to invalidate the original meanings. The meanings of an appreciable number nevertheless, particularly of compound names, do still express the state or character of the places to which they apply. Our military commanders in the North African campaign of the War of 1939–45 learned that to understand the names on the map was to understand the country.

It is of great importance, not merely to the soldier but also to the geographer, to understand the names on the map. Explanatory lists, or **glossaries,** of geographical terms and expressions have therefore been collected from the names on British and foreign maps and charts, and are now regarded as indispensable to geographical study. The words collected come from two somewhat different kinds of source: those from the face of the map or chart, found in the place-names and other written information; and those from the margins, outside the frame of the map, where particulars of scale, date of and authority for survey, drawing and publication, details of price, and explanation of the abbreviations and contractions, conventional signs, colours, stipplings, rulings, etc., used, are given. The elucidation of abbreviations and contractions is particularly important. 'Contr.' (for 'contraction') is an abbreviation; but 'Abbn' (for 'abbreviation') is a contraction. It is now customary to write English contractions which consist of the first and last letters of the word, without a full stop; e.g., *Mr* (for mister) or *St* (for Saint). Many foreign contractions are written and printed in ways unfamiliar to us; and their use in many languages (e.g. Italian) is most inconsistent.

Some of the earliest and best of modern work on geographical glossaries was that of Général Théodore Parmentier, who produced admirable glossaries of German, Arabic, Hungarian, Turkish, the Scandinavian languages and the Romonsch dialects between 1849 and 1896. In 1904 Alexander Knox produced his *Glossary of Geographical and Topographical Terms* (supplementary volume to Stanford's *Compendium of Geography and Travel*, London, 1896–1915) which is still invaluable. Although in part out-of-date on account of reform in the spelling of many languages since it was published, it is still the only general source of information applicable to maps of Asia and of remote regions elsewhere; and the introductory essay contains a useful statement on African languages, particularly of the Bantu and Berber groups. P. Pollacchi continued Parmentier's work in France by issuing a Russian glossary (1907) and an English glossary (1908). Since January, 1943, the War Office has been issuing a series of *Short Glossaries for Use on Foreign Maps*, which already covers about forty languages, while the P.C.G.N. is issuing a series of fuller glossaries for general geographical use. Of the latter, the Modern Greek, Russian, Albanian, Serbo-Croat, Romanian, Thai (Siamese), Turkish and Japanese have already appeared. They greatly facilitate the reading of geographical documents, and the understanding of geographical names, in the respective languages. American official authorities have done similar work, and the American Geographical Society has produced *A German and English Glossary of Geographical Terms* (1950), which is excellent. It is to be hoped that a polyglot glossary of geographical terms now in preparation at the University of Lund (Sweden) will equip geographers with a sharp and valuable tool of trade.[1] UNESCO is keeping the production of all kinds of technical dictionaries, including geographical glossaries, constantly under review.[2]

Geographical glossaries, unfortunately, are not all equally good.[3] Their compilation calls for an exact interpretation of the map in detail, and for an equally exact appreciation of geographical terms in the foreign languages

[1] Kant, E. Preliminärt meddelande om den flerspråkiga geografiska ordboken. *Svensk Geogr. Årsbok*, 1951, pp. 159–167.

[2] Holmstrom, J. E. Report on interlingual scientific dictionaries. UNESCO, *Paris*, 1951.

[3] For examples see *Geogr. J.*, 105 (1945), pp. 64–65; 112 (1948), 229; 114 (1949), pp. 118–119 and 119–120.

concerned and in English. A good glossary becomes a technical supplement to the dictionary of the corresponding language. A geographical glossary is commonly mistaken for a vocabulary of a foreign language for the use of persons travelling in the country to which it applies. That a glossary to maps must, if it is to be of any use, be first of all a list of the very words found on the maps, is not readily clear even to linguists. Some linguists are pained to find obsolete spelling, archaic words, and even words which are not found in dictionaries, in certain glossaries. But the glossaries should give exactly and only what appears on the maps; and map-language often confronts the linguist with the unexpected, for the map produces a host of dialect words, words not in dictionaries, and words used in senses not explained in dictionaries.

Some geographical names have **short forms** for local use (Newcastle, for Newcastle-upon-Tyne, Hull for Kingston-upon-Hull); the City, i.e. of London; the Borough, i.e. of Southwark. A name may even have a contracted written form or forms (Pa and Penna, for Pennsylvania; Worcs for Worcestershire). Other names may have **long forms** for particular purposes, especially for postal distinction (Piaski k. Gostynia, to distinguish it from Piaski k. Lublina, 'k.' being the abbreviation of the Polish word for 'near').

The existence of compound names and of long forms may sometimes lead to doubt as to how much is really the name. It is questionable whether the correct geographical name is 'the Thames' or 'the River Thames'. The proper element, of course, is Thames, and it is permissible to write 'the river Thames'. With such a name as Beverley Brook, on the other hand, most people would agree that 'Brook' is part of the name, because we always say it in full (we never call it 'the Beverley'), and also because it tells us that the feature is not a torrent, a river or a rill. What is essential, and proper to the brook, is the word Beverley, and under that it would be indexed even if we were to call it 'Brook Beverley'.

The test of how to index a compound name, or geographical expression used as a name, will generally reveal how much is essential. It shows up what has to be treated as the 'proper' element. The French observe the helpful custom of joining the parts which constitute this element by hyphens. They call the straits of Dover 'le Pas de Calais', and they index it under Calais; but the department which takes its name from the straits they call 'le département du Pas-de-Calais', and index it under Pas-de-Calais. To index such English names as Lake of the Woods under 'Woods' would be absurd, and we must concede either that the whole expression is proper to the feature, or that it contains no true proper element. With 'Isle of Wight', on the other hand, Wight is a true proper name (Ptolemy's Vectis) and should form the heading for indexing; but because we always say 'Isle of Wight' in full, the whole is a geographical expression used as a name.

In the Borough of Hampstead (London) the following run of street names is used:

Belsize Avenue
Belsize Crescent
Belsize Grove
Belsize Park
Belsize Park Gardens
Belsize Square
Belsize Terrace

Even in Hampstead it is necessary to specify in full which street named Belsize is sought, if one is to avoid being misled. Each of the above expressions is in fact used as a name, and is effective only when used in full. But there are also streets called Belsize in other parts of London than Hampstead. To distinguish them it is necessary to add reference to the postal district, when the expression becomes not a geographical expression used as a name, but something very like a postal address (e.g. Belsize Avenue, London, N. 13). Moreover, these other streets are not necessarily known to the residents of Hampstead.

In general we may therefore say, of geographical expressions used as names, and of long forms of geographical names, that the 'name' is that part of the expression or form which must be used in speech to identify the place or feature concerned for those to whom it is best known. Frenchmen must, except locally, distinguish between Boulogne-sur-Seine and Boulogne-sur-Mer, Germans between Frankfort-am-Main and Frankfort-an-der-Oder, Englishmen between Kingston-upon-Hull and Kingston-upon-Thames; and residents of Hampstead between Belsize Park and Belsize Park Gardens. Nevertheless, the Bristol Avon and the Salisbury Avon; Sydney (New South Wales) and Sydney (Nova Scotia); Richmond (Yorks), Richmond (Surrey) and Richmond (Virginia), show that the rendering of geographical names is an art rather than a science. Much may depend upon the relative prominence of a place or feature under consideration.

Style of presentation is important in the writing of geographical names, though, at least in English, the degree of possible variety is not great. Style is exhibited, in the way in which we connect the words in compound names, and in the use of upper and lower case letters. Between Kingston on Hull, Kingston upon Hull and Kingston-upon-Hull there is room for choice, the different styles representing popular usage and conflicting Departmental practice. There is no general official way of presenting the name. This applies also to Cape Town and Capetown. But between Campbelltown (New South Wales), Campbell Town (Tasmania) and Campbeltown (Scotland) there is no choice. The different written forms are correct for the respective places. It matters little whether we write 'cape Horn' or 'Cape Horn'; 'river Dee' or 'River Dee'; but it might be less appropriate to write 'cape of Good Hope'. What matters is consistency within the limits of the work in hand. In official work, Departmental practice should be followed; in academic work, that of the institution; and in business or private work, the style of an accepted authority (like the Royal Geographical Society) should be taken as a model.

English geographical names. An English geographical name, in the most general terms, is the spoken word or words by which any part of the world or any place in the world is known in the English language. The names of places in England itself constitute a very important part of the language. They are the concern of toponymists, particularly of the English Place-Name Society, and of all Departments of Government which deal with Home Affairs. They are entered on our charts by the Admiralty and on our maps by the Ordnance Survey. The spoken forms of many of them, the live words, now differ considerably from those suggested by the accepted spellings. It is customary to refer to them as 'English **place-names**'. They attract many students, both English and foreign.

We are more interested here, however, in the names used in English for other parts of the world and for places and geographical features outside England. These I shall call 'English geographical names', or, for those who prefer jargon 'English **exonyms**'. They are of interest to all authorities in the United Kingdom and in certain parts of the English-speaking world which are generally concerned with External Affairs, but they are not the particular concern of any one authority. They are of two kinds: names taken into the language from abroad as English knowledge of the world expanded, and English names given to places abroad by English travellers and explorers.

In England the names of many other countries and of many now well-known places and features in other countries have been reduced to an English pronunciation, or to an English spelling, or to both, by constant use. They are anglicized names, or, if you prefer, garbled forms of foreign names. An unfamiliar name makes us hesitate, not from a desire to pronounce it in the correct foreign way so much as through trying to say it in a generally acceptable way that fits into English speech. A foreign name which finds a generally accepted English pronunciation may take a lasting place in our geographical vocabulary. Even the name of one of the nearest of foreign places, Boulogne-sur-Mer, many of us pronounce 'booloin', which is far from being French. It is interesting to note, nevertheless, that our present pronunciation of some foreign names may preserve something of their earlier pronunciation in the languages to which they belong. Our Paris, for instance, is in one way nearer to old French than to modern French, in which the *s* is sounded only in certain conditions before a vowel, as in the adjective 'parisien'. Just as we may garble the pronunciation of a foreign name when accepting it into English, we may also modify its spelling. English is written without diacritical marks, and we fail accordingly to respect those of other languages when we write foreign names in English context. We never take the trouble to write 'Rhône' in English context, where 'Rhone' will always do. Nobody writes Panamá or Pará in English.

Names associated with certain earthworks in Scandinavia are thought to be relics of the speech of prehistoric times. We may assume that knowledge of the names of some countries and places abroad was brought into Britain by its earliest invaders. We may also assume that anglicization of the more important Cornish, Welsh, Scottish and Irish names began here in remote times. Anglo-Saxon historical writings would undoubtedly provide a small stock of anglicized names of countries and places remote from Britain. From the time of the First Crusade foreign geographical names have probably been invading the English language in steadily increasing numbers. Settled mediæval British usage is displayed by Chaucer, who was fairly consistent in his spelling of foreign names. His standing as a hydrographer, and his position as Comptroller of Customs in the Port of London vested him with official authority. He wrote of Beme (Bohemia), Portyngale (Portugal) and Surriye (Syria, not Surrey), and indeed, spelled only two foreign names as they are spelt to-day: Jerusalem and Rome.

With the development of English cartography and the expansion of our maritime interests in the fifteenth century, foreign geographical names, recorded mainly by foreign adventurers and missionaries, began, so to speak, to breed; and by the time the limits of the world were understood our need for precision in writing the names became fairly evident. Names of foreign places as Marlowe and Milton (also a Government official) wrote them, represent, besides reform in orthography, successive stages in correction. The introduction of foreign names particularly by war, has continued right down to our own times. Marlborough's campaigns gave us Oudenarde, and the Crimean war gave us Sebastopol. They represent strata of loan-words.

Foreign names which have so established themselves in anglicized form in spoken English as to be in frequent and general use may justly be regarded as English names. They have received very little study as names, and may be more numerous than we suppose, because some of them are hard to recognize. I have suggested that, Genua being Latin, Genova Italian, and Gènes French, Genoa is probably English; and that, Genava being Latin (or latinized Keltic), Ginevra Italian, and Genève

French, Geneva is also probably English.[1] But what is Leghorn?[2] The meaning and history of the corresponding Italian name, Livorno, may appropriately be left to Italian toponymists, but the establishment of the form Leghorn in English is for Englishmen to explain.

Our anglicized names all have their accepted written forms. For the innumerable foreign names which appear in print in English but are never, or hardly ever heard spoken, we have written forms in embarrassing variety; and for hosts of foreign names of which we never even hear, our cartographers have to provide written forms.

Foreign names which have been accepted into English may have been borrowed directly or indirectly. Many places in parts of the world where English is not spoken are known to us not by English distortions, oral or graphic, of the local names, but by names which, though clearly not English, are not used at the places to which we apply them. The Crusaders brought us a name for Cairo not from the Saracens, who called it القاهرة (al Qāhira), but from the Venetians and the Genoese. We call the capital of Denmark, Copenhagen, but the Danes call it København. The Danish name is particularly hard for us to pronounce and to print, but its German form, Copenhagen, we can manage with ease, though we pronounce it imperfectly. These are not perhaps the most important reasons why we have accepted it, but we found it very convenient to do so. Such a name, taken into English from a source foreign to the place to which it refers, or from a former state of the language spoken there, is a **'conventional' name,** or, more precisely, a 'conventionally English name'.

Conventional names are widely used in English in circles of special interest. Many of them are accepted in theological,

[1] *Scottish Geogr. Mag.*, 63 (1947), p. 3.

[2] *See* Smyth, W. H. The Mediterranean, *London*, 1854, p. 409; and correspondence in *The Times*, 8th, 13th, 14th and 21st July, 1944.

classical and historical studies of the Old World. The iron-ore trade has its conventional forms of reference to some of its ports and deposits (e.g. Bona for Bône); and the timber-trade provides an example in the name 'spruce', which comes from the Polish expression 'z Prus', meaning 'from Prussia'. The recognition of some conventional names is not necessarily a simple matter. Salonica, for instance, is not, as is commonly supposed, a faulty spelling of Salonika, for the latter form has no right to existence. Salonica is a true conventional name, taken from the Spanish-speaking Jews of Thessaloniki. Again, Cologne is not conventional because it is French for Köln, but merely because it is French, both Köln and Cologne being direct descendants from the Latin Colonia, one by the German, the other by the French line.

Conventionally English names are perhaps even more important than anglicized geographical names, for, coming to us indirectly, from other tongues and other times, they have a degree of international currency, and are, in fact, not only English but also Occidental. Little has so far been done towards their recognition and collection, but, like anglicized names, they await geographical and linguistic study.

Mature use of conventionally English names was exhibited by some of the later English men of letters (Gibbon, Ruskin, etc.), and is reflected by many names in the Concordance to the Bible, in the indexes to such standard works as *Hakluyt's Voyages* (12 vols., Glasgow, 1903–1905), *Purchas his Piligrimes* (20 vols., Glasgow, 1905–1907) and Suess's *The Face of the Earth* (5 vols., Oxford, 1904–1924). It is found in W. H. Smyth's *The Mediterranean* (London, 1854), in Longman's *Gazetteer of the World* edited by G. G. Chisholm (edition of 1905), Chisholm's *Handbook of Commercial Geography* (editions before 1914), the *List of Countries* issued by H.M. Stationery Office (London, 1925), *Lloyd's Book of Ports and Shipping Places*

(London, 1944) and particularly on the maps in Bartholomew's *Citizen's Atlas of the World* (edition of 1912).

English geographical names of the second kind, that is, English names given to places abroad by English travellers and explorers, are exemplified by the name Clarence. One has only to think of Clarence strait, between Qishm and the mainland of Persia (but which has no Persian name), or of the Clarence river in New South Wales, to appreciate its importance in the catalogue of English geographical names, but, like many other names, it is difficult to explain. It commemorates, as a rule, King William IV when Duke of Clarence. But where is Clarence? There is no place so called in England except Port Clarence on the Tees. I am told, by the College of Arms, that the name relates to Clare in Suffolk, and that there is no foundation whatever for the belief that it refers to a supposed ancient duchy of Chiarenza or Glarentza in the Peloponnese. Nor, I believe, does it relate to County Clare in Ireland, no Earl of Clare ever having been also a Duke of Clarence. The derivation from Clare in Suffolk is possibly through a Latin genitive (Clarensis), as Isaac Taylor suggested.

The number of anglicized, conventionally English, and exported English names for places abroad is probably small, considered country by country, and they are of two grades: those known to all educated Englishmen, and those known only to Englishmen of special knowledge (sailors, geologists, archæologists, historians, etc.). A name such as 'the Gulf of Danzig' needs no explanation. Everybody knows or ought to know what and where the gulf is. But the name 'Hel point' (also a feature on the coast of Poland but one without a Polish name) is known only to English sailors who have been in Polish waters.

English geographical names as a class are somewhat misunderstood. Many would abolish them and employ the foreign official names instead. In June, 1939, the Siamese

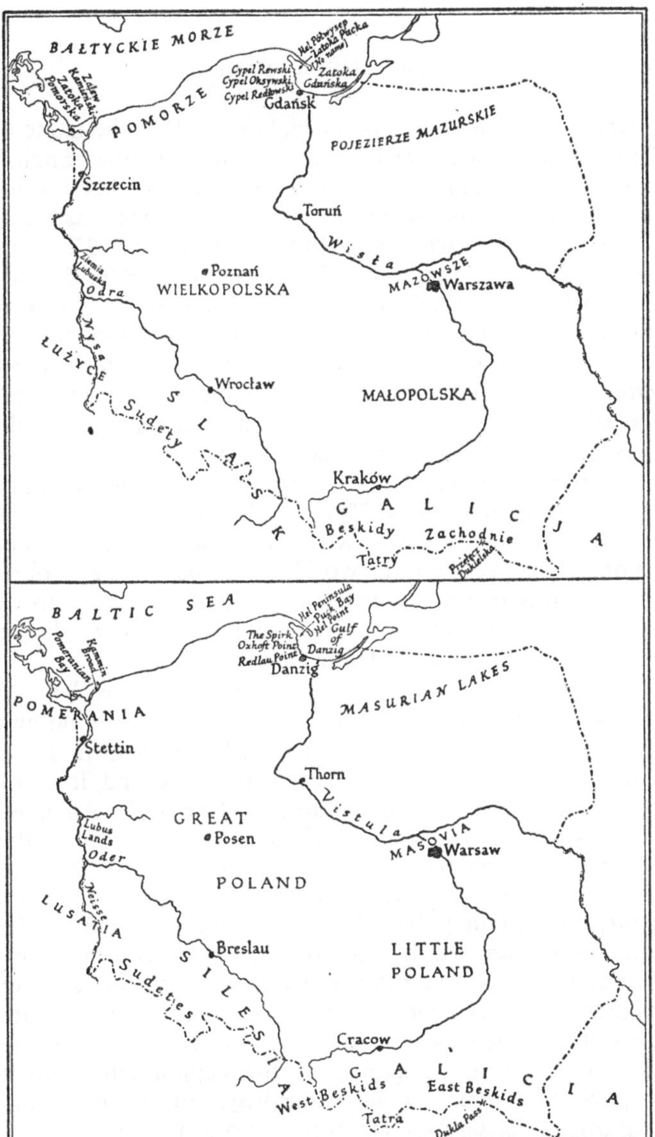

FIG. 2.—*Polish names of Polish places* (top) *and the corresponding English and conventionally English names* (bottom).

Government announced by official decree that the name of Siam would thenceforth be Thailand 'in the English language'. Sir John Squire protested[1] against this curious announcement, and wrote: 'When that change was made I asked a Siamese why on earth they had done it. His reply was that they thought the change would please Europe, and particularly England. I said: "Look here, my dear fellow, all Europe since Elizabeth's day has known your country as Siam; . . ." ' Unfortunately, British authorities acquiesced in the change, and the bastard name Thailand has gained currency as a postal name. But that we should renounce our living name 'Siam' in favour of one which the Siamese themselves do not use (they call their country Prathet Thai or Muang Thai) is, I hope, unlikely while English is a living tongue. Moreover, the Siamese are not the only Thai people in the world. It was no more appropriate for the Siamese to decide what their country is to be called in our language than it would be for us to decide what anything is to be called in Siamese.

The utility of English geographical names is beyond question. In the first place, they come at once to mind and fit naturally into our speech, for we know them and can pronounce them without hesitation. In the second, it is not convenient to refer to the Danube, the Nile or the Amazon, in English context, by any of the names in use in the countries through which they flow. In the third, to have to choose between several official names for a place or feature abroad may be offensive to one or some of the countries concerned. In the fourth, anglicized and conventionally English names may survive in English use after changes of government or even of sovereignty abroad. Our well-tried names for the oceans and continents, of the seas, countries and regions, of the capital cities, great sea-ports and international water-ways, of the mountain

[1] *Illustrated London News* ,198 (1941), p. 372, and 202 (1943), p. 548.

ranges, greater lakes and great rivers, and of all those features and places which have become symbols in the English imagination, may continue to live in our own daily speech. That they will survive in countries in which English was formerly the official language is, however, doubtful. The anglicized Benares, Cawnpore and Lucknow are already Banāras, Kanpūr and Lakhnū when written in Roman letters in India.

English, which developed as the language of the people of England (that is, of the English people), is now, however, the mother-tongue of all people of British descent in Her Majesty's Dominions and is the national language of the United States of America. It is the official language of the United Kingdom and is officially used (though not everywhere as a first language) in the new British Empire. It is developing regional characteristics in several parts of the English-speaking world, and we recognize certain words, expressions and habits of speech as American English, Australian English, and so on. Americans do not say the word Paris just as we do. It has, within the English language, both American and English spoken forms.

The names of numerous places in other countries where English is the tongue of all or of a considerable proportion of the people, are of English origin, and, strictly speaking, are to that extent English names. But they, and distortions of many native names of places in those countries, are an important part of the vocabulary of the variety of English now spoken there. It will be convenient to refer to them as 'American place-names', 'New Zealand place-names', etc. Launceston (Lawns-ton or Lahns-ton) is in Cornwall, but Launceston (Lawn-ces-ton) is in Tasmania. It must not be forgotten, however, that there are French geographical names in Canada, and that they are also a part of the sum of Canadian place-names; that there is a legitimate study of Scottish place-names, but that 'Scottish geographical names' do not exist. Gaelic geographical names, on the contrary, belong to reality, and those of places in Scotland form a large part of the sum of Scottish place-names.

If English geographical names as a whole are not the concern of any one authority, neither have they attracted many students. R. S. Charnock (active 1859–1894) and Isaac Taylor (1829–1901) merely glanced at them. Some names will be included in the place-name studies now proceeding in English-speaking countries overseas, where they can hardly be appreciated with regard to the English of England. M. J. C. Meiklejohn's book *The Place-Names of the English People at Home and Overseas* (London, 1929) is a slight work and essentially a place-name study.

English has suffered no more neglect in this respect than other languages. I know of only one German paper which deals directly with German names of places abroad.[1] There is a paper on Arabic names by C. A. Nallino [2] and a small work [3] which I have not seen by A. J. Schreuder on Dutch names abroad. Portuguese names on the west coast of Africa have recently been collected by A. Texeira da Mota,[4] and small lists of Chinese and Japanese names of places outside China [5] and Japan [6] exist.

In the meagre literature of English geographical names an anonymous paper (actually by G. R. Crone)[7] on 'Names of countries used in other countries' is important, and

[1] Seager, R. Deutsche Namen nichtdeutscher Orte. *Vjhft. geogr. Unterr.*, 2 (1903), pp. 156–160.

[2] Les noms géographiques du monde musulman dans les publications arabes modernes. *B. Soc. géogr., Caire*, 7 (1907), pp. 1–17.

[3] Nederlandsche plaatsnamen buiten de landsgrenzen. *Middelburg*, 1939.

[4] Centro de Estudos da Guiné portuguesa, No. 14. *Bissau*, 1950.

[5] List of countries. *Chinese Maritime Customs, Statist. Ser., 'The Trade of China'*, *passim*.

[6] 外國地名及人名書キ方及稱ヘ方調査表 (List of foreign proper names). Suppt. to *J. Geogr.* (*Tokyo*), 14 (1902), No. 168.

[7] *Geogr. J.*, 85 (1935), pp. 458–462, and 86 (1935), pp. 179 and 558.

several studies by the Swedish toponymist Gösta Langenfelt are exemplary. In his *Toponymics or Derivations from Local Names in English* (Uppsala, 1920) he has assembled a great amount of information about words derived from English geographical names, which will help in the recognition of these names; and his paper 'Sweden; a monograph on the names of a country' must rank as one of the most thorough studies of its kind.[1] If we are to know and understand our English geographical names, much more work like that of Langenfelt must be done. We should collect our spoken forms in the International Phonetic Alphabet. Part of this work has already been done by Professor Daniel Jones [2] and by the British Broadcasting Corporation.[3] We should then collect our written forms from English literature and maps, and it would be most instructive to begin with our forms of Welsh and Irish names.

The following tree displays the relationships between the different kinds of English geographical names as I have treated them here:

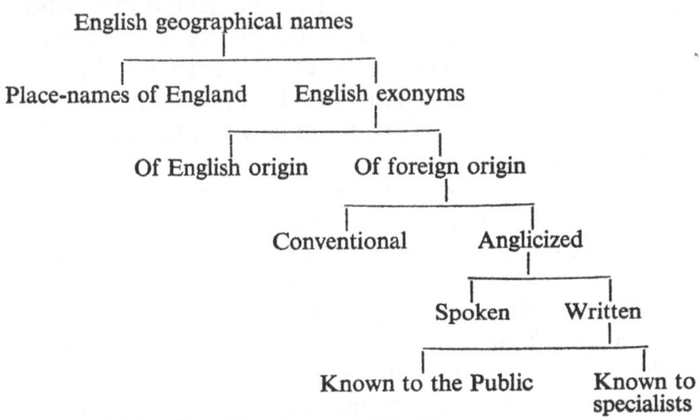

[1] *Neuphilol. Mitt.*, 33 (1932), pp. 92–121.
[2] An English Pronouncing Dictionary, 9th ed., *London*, 1948.
[3] Broadcast English, i–vii, *London*, 1928–1939.

Occidental Names. There is a stock of geographical names (small, I believe) in use in the languages of Europe, derived from such sources as the Christian and Hebrew scriptures, Greek and Latin literature, and the Ptolemaic geography of the Renaissance, which it may be profitable to recognize. The names are of oceans and seas, countries, regions, prominent geographical features, places which are or once were of religious, commercial or strategic importance, and of peoples and languages; and they have assumed the same, or similar forms in several languages. The form Vienna, for instance, is used in English and Italian, Viena in Spanish, and Vienne in French, while the German Wien looks a more distant relative than it sounds. Only the Hungarians have a really different name for the place: Bécs.

These names are the descendants of the geographical nomenclature of the Old World. The ancients recognized the geographical entity of certain prominent features of their world, and expressed it in the names they gave them. It has not always followed that the peoples now living in the neighbourhood of such features understand their wholeness and express it in a single name. There is still a use, in English, for such river-names as Oxus, Jaxartes and Araxes, and for such mountain-names as Paropamisus.

There is also a stock of groups of newer names (very small, I believe), in use in European languages, in which each group represents a set of translations from an original. The Cabo da Boa Esperança of the Portuguese for instance, has become the Cape of Good Hope, Kaap de Goede Hoop, Cap de Bonne Espérance, Kap der guten Hoffnung, etc. The Chinese call it 大 浪 山 (Ta Lang Shan) 'the great wave mountain'[1] which expresses an entirely different idea.

These sets of names differing slightly in form, and sets of forms of identical meaning, deserve study. Those of the

[1] *Geogr. J.*, 52 (1918), pp. 367–385, and 53 (1919), pp. 19–30.

first kind may be of interest to Oriental peoples in their efforts to assimilate Occidental names into their own languages. Those of the second kind, which belong mainly to the period of European expansion, may contribute little or nothing to Oriental convenience but they may prove to be of importance in establishing regularity in Occidental international usage. The two kinds, taken together, represent the main source from which the United Nations have drawn their list of approved names in English, French and Spanish, of statistical areas; and from which the International Civil Aviation Organization has drawn its list of approved names, in English and French, of countries and of international air-ports. According to *Petermanns Mittheilungen*, 83 (1938), p. 277, the congress of the International Geographical Union held at Amsterdam in 1938 recommended the use of these 'international' names in brackets after 'local' names; and as long ago as 1933, Gösta Langenfelt urged their use, in Sweden, for international dealings. Many of them belong to the category of 'conventionally English' geographical names already considered above.

Foreign Names. All geographical names in languages other than English are, in the narrow sense, foreign geographical names. But we hardly regard Welsh or Gaelic names as foreign; nor can we exclude numerous American and Canadian names from the English language. Neither are we likely to be greatly concerned with what Estonians call places outside Estonia. In practice, we have to deal with foreign local names in the official languages of foreign sovereign states (e.g. Kharhov, Russian Federal official), in the subordinate languages of such states (e.g. Kharkiv, Ukrainian official for the same place), and with names in the native languages of states or territories under protected or colonial rule (e.g. Beqa, Fijian; but written Mbengga if rendered in English form). They are of importance to all authorities and organizations interested in External Affairs, and are of special interest to H.M. Services and Departments of State.

Foreign habits of naming are not necessarily the same as ours. Greek sailors, for example, have names for the prominent features of Greek maritime geography, even to the winds of the Greek region, which are different from those used by Greek landsmen. It is not possible to say, of such alternatives, that one is right the other wrong. In parts of New Guinea people may have names for all surrounding villages but none for their own.

Many supposedly native names of Melanesian islands are merely the native names for parts of islands. Santa Maria, in the Banks islands, for example, is commonly supposed by us to be called Gaua by the natives; but Gaua, to them is merely the name of the north-eastern part of the island in the language of that part of the island.

The Polynesians, a more advanced people, have names for all their islands, but may not have names for groups of islands. I doubt that there was any name in old Tahitian corresponding to 'Society Islands', or in old Hawaiian corresponding to 'Sandwich Islands'.

Peoples given to generalization of statement will have names for many things not named by less advanced peoples. The Greeks recognize the structural relationships between a certain mountain-chain, peninsula and promontory by calling them Keravnia Ori, Akrokeravnia Ori and Akrotirion Akrokeravniou, respectively; but the Albanians, in whose country they now lie, have no corresponding sequence of names, though they have names for every pass and summit of the structure. The name 'North-Albanian Alps' is a useful English regional name; but the highlands of northern Albania and neighbouring Yugoslavia are not recognized as a unit by Albanians and have no Albanian name.

There is now a serious need for an extended study of the ways in which geographical names are used in certain foreign countries, which would supersede Isaac Taylor's admirable essays on French, German, Slavonic, Hungarian, Turkish, Indian, etc., names.[1]

Foreign geographical interpretation of a region may be very different from our own. Notwithstanding the direct relationship between the names Biscay and Vizcaya, the English 'Bay of Biscay' is a far larger sheet of water than the Spanish 'Golfo de Vizcaya' or the French 'Golfe de Gascogne'. The correct identification of the lesser chains and ridges of parts of the Carpathians by the names used for them in English, Polish, Czech, German and Russian is exceedingly difficult because the several nations have analysed the mountain-system in different ways. It would take onomastic study of what we call 'the Beskid mountains' to clarify the nomenclature of maps of this region.

The name Romania is a remarkable example of a name which has been used for different things at different times, and is now used in different ways by different authorities.[2] Meyer-Lübke maintained that the old Roman Empire of Latin-speaking world was called Romania during the Middle Ages. He should perhaps have added 'by Europeans', for the lands of Christendom were called the lands of Rūm by the Saracens. The Muhammadan traveller Ibn Jubayir shipped from an Andalusian port late in the twelfth century, and passed, on his way to the East, between the 'islands of Romania', identified as the Archipelago. After the final defeat of the Eastern Roman Empire in 1204, the Franks established the short-lived Empire of Romania, with its capital at Constantinople; and the name Romania seems to have lingered, as that of lands at the head of the

[1] Names and Their Histories. *London*, 1896.
[2] *Geogr. J.*, 108 (1946), pp. 268–270.

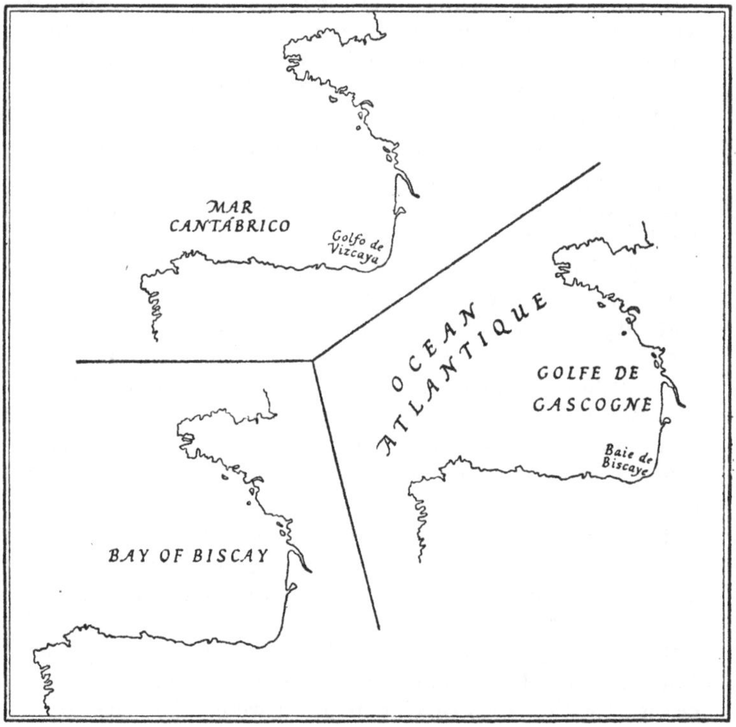

FIG. 3.—*Names of the Bay of Biscay: Spanish* (top), *English* (bottom left) *and French* (bottom right). *French practice is not consistent. Many French maps and charts show the Golfe de Gascogne, as indicated above, but few show the Baie de Biscaye.*

Aegean, at least until the establishment of Turkish power in Europe was generally acknowledged. Older English atlases, from Speed (1626) through Bowles (1708) and Senex (1721) to Kitchen (1782) attached the name Romania to a vaguely delimited region covering parts of Thrace and Macedonia. The town of Nauplia (Navplion) in southern

Greece is still conventionally known to Italians as Napoli di Romania.

The descendants of the old Latin-speaking tribes on both banks of the Danube had come to style themselves as Romeni, Romani, Rumeni or Aromuni from very early times. Those of Moldavia, Wallachia and Transylvania called their country 'ţara românească', 'the Romanian country'. In 1859 the Representative Councils of the Principalities of Moldavia and Wallachia voted unanimously for union into a single state to be called România. The union was ratified by the Powers in 1861; but not until 1882, it seems, did the name Romania again begin to appear in English atlases (Philips's Imperial Atlas), though in a different form (Rumania) and in a different place from where it had appeared in seventeenth and eighteenth century atlases. It is now officially spelt Romînia.

A third use of the name is that of modern linguists, who apply it much as it was used in the Middle Ages, to denote the whole area over which the Romance languages are spoken. A journal called *Romania* devoted to the study of these languages was founded in Paris in 1882. Linguists are therefore constrained to refer to the modern state as Rumania or Roumania, to distinguish it from their Romania, in spite of the Romanian official spelling of the name.

Names of the same or of similar meaning are not necessarily applied to the same things in different languages. Our North Sea is Vester Havet (the Western Ocean) of the Danes, and our Baltic is their Ostersöen (the Eastern Sea). The Windward and Leeward Islands represent a geographically illogical division of the Islas de Barlovento of the Spaniards (or Eilanden Boven den Wind of the Dutch), our Leeward Islands being no more 'leeward' than the Windward Islands; and for the Islas de Sotavento of the Spaniards (or Eilanden Beneden den Wind of the Dutch) we have no name. Nor are our lesser Antilles (the Pequeñas Antillas of the Spaniards) now the same as the Kleine Antillen of the Dutch.[1]

On pages 12 and 13 of the twelfth edition of *The Channel Pilot, Part 1*, the Admiralty used the forms Dover Strait, Dover strait, and Strait of Dover (an inconsistency not apparent in the current (13th) edition) for what our newspapers invariably call the Straits of Dover. It is therefore not astonishing to find that, in a given foreign country, names may be rendered in different forms by different authorities, and even by the same authority at different times. Even in a country geographically so well ordered as France, a name may be written in one way by the Post Office but in another by the Railway Administration. The amount of disagreement between authorities, both official and unofficial, concerning the forms of geographical names is appreciated only by those who have to work closely with the large-scale maps and charts, atlases, books, reports, and articles in journals, from which the choice of an appropriate form for a foreign name has so often to be made. Nor is the disagreement confined to the names of well-known places and features, so many of which have English or conventional names. The countless names of obscure places and minor features, any of which may become important on a War Office map or Admiralty chart, are rendered, even by highly organized countries, with an inconsistency of which few but cartographers are aware.

Foreign names, by definition, belong to all the foreign languages of the world. But foreign languages are written in a great variety of ways, many of which differ radically from the English alphabetical system of writing. Names written abroad in unfamiliar scripts must therefore by some

[1] Jeffreys, T. The West-India Atlas, *London*, 1775 (ed. of 1810).

means be reconciled with our English alphabet if they are to appear on our maps and charts. And herein lies the difficulty of writing them in ways suitable for English use. Finally, many foreign names contain technical (mainly geographical) terms, which must be explained if the names are to be properly understood. (*See* pp. 11–14.)

CHAPTER III

SYSTEMS OF WRITING

The number of speech-sounds, and the number of ways of uttering and modifying them, is so large that no language exhibits them all, no individual can use them all, and no simple system of writing can represent them all. Sounds in European languages are made by the release of breath, but in many African languages there are sounds made by drawing air into the mouth, that is, by suction. They are called implosive sounds. The only sounds known to us which resemble them are that of a kiss and those of certain clicks of the tongue. In many African and Asian languages the meaning of words may depend upon the tone of voice in which they are uttered. Examples of the use of tone to help to distinguish meaning are numerous even in English, and are illustrated by the difference in sound, in unemphatic statements, between *present* (noun or adverb) and *present* (verb); or by the different ways in which we say 'a black bird' and 'a blackbird'. In other languages, particularly of the Turkic group, much depends upon the length and even on the harmony of the constituent sounds. In still others, certain stops to sound are of importance. Our final sounds of *p*, *k*, and *t* in such words as top, lock, pot, are stops which have to be exploded before the next sound can be made. A stop in the throat, exploded by the sound we make in a weak cough, and called the glottal stop, is important in some languages, as in spoken Danish. In Tongan (Polynesia), *ua*, *'ua*, *u'a*, and *'u'a* all sound different and have different meanings. Tone, length, and the glottal stop are not easily indicated in simple alphabetical script, but are essential to the clear written representation of some languages.

The unit of sound in a language is now regarded as the **phoneme,** and it may be defined as that range within which a sound may vary without affecting the meaning of words in which it occurs. For instance, it does not matter how we make the sound of 'p' in English, as our different ways of saying it do not affect meaning; but we cannot say 'fool' when we mean 'full', although the vowels differ only in length. When we learn French we must be careful not to say 'p' with the puffing sound often heard in English, not because it would affect meaning, but merely because it would not be French; but, should we attempt certain Indian languages, it would be necessary to master two ways of saying 'p' clearly if we were to be understood at all. Two of the several possible ways of saying 'l' are heard in the word 'little', but in English they do not affect meaning. In Russian they do affect it. In other words, one l-phoneme is enough in English but two are needed in Russian. In modern Greek there are fewer phonemes than in English, but in Polish there are more.

The number of known living languages is enormous and amounts to hundreds in India alone. Concerning their origins, affinities, and classification there has been much discussion and there is much difference of opinion. The best source of information is *Les Langues du Monde* by A. Meillet and Marcel Cohen (Paris, new ed. 1952).

Systems for the recording of speech in writing have been in use in some parts of the world from remote times, and the number of languages which have not yet been reduced to writing is probably small. It does not necessarily follow that a system of writing provided by scholars for an obscure language is in use amongst native speakers, but it is true that there are very few languages which cannot be satisfactorily written in one way or another.

The systems of writing in civilized use are of three kinds only; those in which the phonemes of a language are, or originally were, represented by distinct **letters** (alphabetical systems); those in which the syllables of a language are represented by corresponding **symbols** (syllabic systems); and those in which whole words are represented by unique **characters** (logographic or ideographic systems). Combinations of these systems are not only possible but are in actual use in Oriental countries. A system established in use for a particular language may not suit the language, but we have to deal with it and make the best of it. The Arabic alphabet, for instance, is a wonderful product of human ingenuity and a thing to admire; but it is not well suited to the writing of Persian though it has been so used for centuries.

The spoken word is alive, and subject to relatively rapid change. Change in the written word is slow. Once a word has assumed a generally accepted written form, that form becomes a visual symbol, seen as a whole, not letter by letter, and any attempt to change it is opposed by those who shrink from the mental discomfort, the delay in reading, caused by learning to recognize the new symbol. Languages too may change rapidly, but established systems of writing do not, and a system which may have been admirable in the beginning may become clumsy with time. Modern Greek is still written like ancient Greek, and their old, elaborate spelling no longer reflects the pronunciation of Greek words. Generally speaking, however, the established systems of writing are well accommodated to the needs of the groups of languages for which they have been developed.

Those interested in the appearance of foreign alphabets and scripts will find hundreds of examples in *The Gospel in Many Tongues* (British and Foreign Bible Society, London, 1937). A limited choice of foreign alphabets and scripts is considered in more detail in *Alphabete orientalischer und occidentalischer Sprachen* by Friedrich Ballhorn (12th edition, Nürnberg, 1880) and in *Alphabets of Foreign Languages* by Lord Edward Gleichen and John H. Reynolds

(2nd edition, reprinted, London, 1951). They are best displayed in the standard grammars and dictionaries of foreign languages. In further consideration of the subject I shall refer, for the sake of clarity, to alphabetical *letters*, syllabic *symbols* and to logographic *characters*.

Alphabetical script. The true alphabet is one in which the vowels and consonants are all and always represented by distinct letters. The ideal alphabet would be a phonetic alphabet, that is, one in which each phoneme of the language to be written would be represented by a single sign or letter. Very few alphabets fulfil these conditions, but it is probable that most of them, when originally developed, approached them closely. The history of the alphabet was exhaustively studied by Isaac Taylor (*The Alphabet*, 2 vols., London, 1883). Much of his work has been superseded, but the subject is adequately treated in the standard encylopædias and there is no need to consider it further here.

In certain alphabets the letters assume different forms according to their positions in words. We are accustomed in English to the use to upper and lower case letters (commonly called capitals and small letters) for different purposes, and the first words of our sentences all begin with upper case letters. In German, all nouns begin with such letters, no matter what their positions in sentences. In the Arabic alphabet the forms of the letters depend upon whether they stand alone, or at the beginning, within, or at the end of words (i.e. are isolated, initial, medial or final).

Modern printed forms of letters were developed from their earlier, hand-written forms; and in handwriting numerous conventions arose for joining letters and for abbreviating and contracting words. Very few abbreviations or contractions are in regular modern use in running printed prose, but more joined letters, or ligatures, are in use than the average reader is aware. In some founts of type *ffi*, for instance, is cast as one piece of type. In Arabic the ligatures between certain consonants are numerous and have to be learnt.

Alphabets in which the letters themselves are of the same kind as our English letters are called **Roman alphabets** (*cf.* Roman numerals), because the common forms of the letters are derived from the classical lettering of ancient Rome. It is not quite correct to call them 'Latin alphabets', for the Latin alphabet is now only one of a class, and it differs from other Roman alphabets in several respects, having no letter *k*, for instance. Forms of the Roman alphabet have come into general use throughout continental Europe (except in Greece, Russia, Bulgaria, and part of Yugoslavia), throughout the Americas, Australia, and New Zealand; and are in official use throughout Africa (except in Egypt, Libya, and Abyssinia), and in India, part of Indochina, Indonesia, the Philippines, and Oceania.

The English form of the Roman alphabet is one of the simplest, for the bare letters suffice, without any of the diacritical marks used in foreign forms to indicate modification of the sounds normally carried by the letters. In French, *e* normally represents a neutral sound like that of the unstressed *a* in the phrase 'a bird'; but French *é* indicates a sound like *a* in English 'able'; and *è* indicates one like *ai* in English 'fair'. Unfortunately we now have a number of ways of representing some sounds in English. C*a*ne, r*ai*n and r*ei*n all rhyme. We use some letters, particularly those representing vowels, for sounds usually represented by other letters in continental Europe. For the sound of *a* in 'cane' an Italian or a Spaniard would write *e*; for the *ee* in 'been' he would write *i*. For our word 'fine' a Frenchman would write 'faine', and he would be phonetically right, because some of our so-called vowels are really diphthongs. Some, too, are uncommon and constitute phonemes inadequately represented in our alphabet. Our sound of *a* in hat, of *i* in sit and of *u* in but are rare in

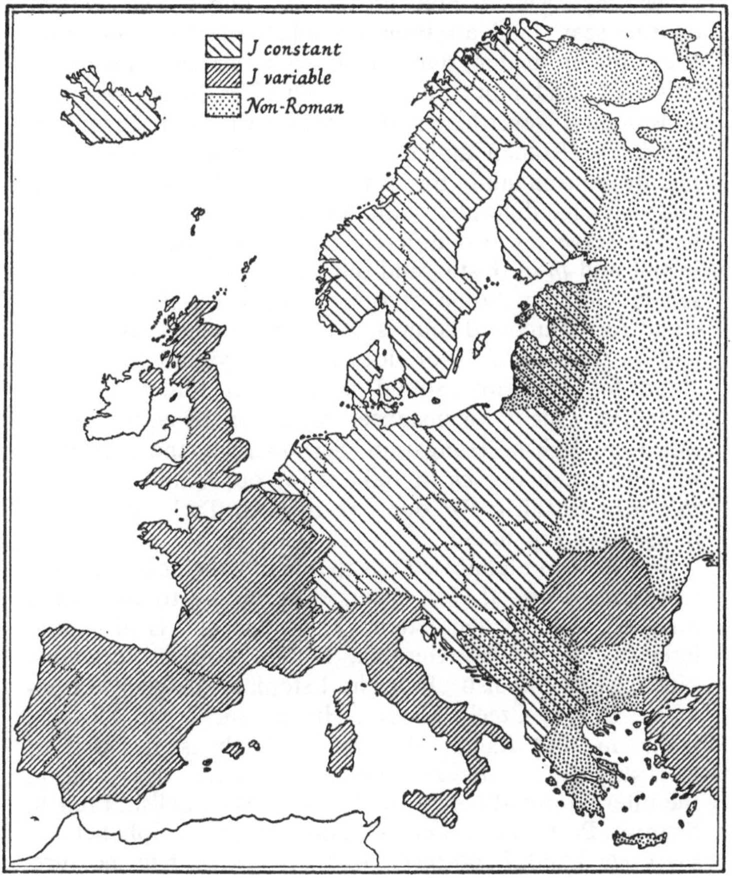

FIG. 4.—*The Roman alphabets, classed according to the value of* j. *The regions throughout which* j *has the value of an English y-before-vowels is also, in great part, the regions in which diacritical marks are commonly used on other letters. In the south-western region* j *carries differen values in different languages.*

other languages. A Frenchman has trouble in saying 'sit on this seat'; and an Italian would write 'bat' for 'but'. We are in better agreement with western European countries, and more consistent ourselves, in representing consonants. Nevertheless, in common with other countries which use Roman alphabets, we represent some sounds, mainly consonantal, by combinations of letters. Our *th* and *sh* (called digraphs) stand for simple sounds; *th*, indeed, has to serve for two different simple sounds, those of *th* in this and *th* in thistle; and our *sh* is written *ch* in French, *sch* in German. Some of our symbols, on the contrary, stand for compound sounds, *x* being the equivalent of *ks*, and *ch* that of *tsh*. Nor does a given symbol necessarily stand for the same sound in different foreign Roman alphabets. Our *c*, for instance, stands for the equivalent of our *ts* in Polish, Hungarian, Serbo-Croat, and Albanian; for our *k* and our *ch* in Italian; for our *k* and our *th* in *thistle* in Spanish; for our *j* in modern Turkish; for our *th* in *this* in Fijian, etc.

The Roman alphabets are capable of rough classification in several ways, as, for instance, according to the values assigned to particular letters. The value of *j* is of special interest, for, in the German, Dutch, Danish, Icelandic, Norwegian, Swedish, Finnish, Estonian, Latvian, Lithuanian, Polish, Czech, Slovak, Hungarian, Croat, Slovene, Albanian, and Italian alphabets it is the same, that of English *y before vowels.* And it also carries this value in the International Phonetic alphabet. In the English, French, Spanish, Portuguese, Turkish, and Romanian alphabets on the other hand, it corresponds to *g* in gem, *s* in measure, or *ch* in loch, according to the language represented. It therefore seems evident that it should carry the value of English initial 'y before vowels' in a possible international Roman alphabet which I shall discuss below.

The Roman alphabets may also be grouped, though not

scientifically, according to their diacritical marks. Here in Great Britain we may regard ourselves as being flanked by zones of foreign Roman alphabets. The Dutch in the North, and the Italian and Spanish in the South, like our own alphabet, are almost free from diacritical marks. The German, Danish, Finnish, and Estonian carry a few such marks; the Portuguese, French, Norwegian, Swedish, and Icelandic display them more liberally; and in the Latvian, Lithuanian, Polish, Czech, Slovak, Hungarian, Romanian, and Turkish alphabets they are numerous.

The Irish alphabet, now being revived in Ireland, is an antique Roman alphabet based on old manuscript forms of the letters. Another antique Roman alphabet of considerable importance is the so-called Gothic alphabet which was formerly in general use in all countries surrounding the Baltic, as well as in parts of the old Austrian Empire. It is better called 'Fraktur', as in Germany, where it is still in use, though no longer in general use. Fraktur is perhaps better regarded as a type-face than as a distinct alphabet, but knowledge of it is indispensable for the reading of German, and for the intelligent use of Scandinavian documents from the recent past.

Non-Roman alphabets are not legible by those who know nothing of the languages for which they are used. In Greece an alphabet closely related to the Roman has been is use for about 3,000 years, but it cannot be read by those who know nothing of Greek. In Bulgaria, part of Yugoslavia, and Russia, forms of an alphabet called the Cyrillic, believed to have been given to the Slavs in the ninth century by St. Cyril, are used. Amplified forms of the Cyrillic alphabet are being developed for application to non-Russian languages in territories of the Union of Soviet Socialist Republics, particularly in Russian central Asia, and are likely to become important. Ancient alphabets survive in local official use in the autonomous republics of

Georgia and Armenia. In Israel the Hebrew alphabet has been officially revived.

Throughout the Muhammadan world the Arabic alphabet, that of the Qurān, is the official medium of the independent countries (Libya, Egypt, the Sudan, the states of Arabia, Jordan, the Lebanon, Syria, 'Iraq, Persia, Afghanistan, and western Pakistan); and it is in subordinate use in Morocco, Algeria, Tunis, French West Africa, French Equatorial Africa, northern Nigeria, the Aden Protectorate, the Somalilands, Zanzibar, and Malaya; and it is still used by Turkic and Iranian peoples in parts of Russian central Asia and China. The extent to which the Mongolian alphabet (a very distinctive one) remains in use amongst Mongols in Mongolia, Russia, and China is very doubtful. Of the few true alphabets in use in South-East Asia the most important is that of Siam, Cambodia, and Laos.

Mixed alphabets. There are a few partly Roman alphabets in use which are classed as mixed alphabets. The Icelandic, Faroese, and Maltese alphabets each contain one or two non-Roman letters; and the International African Alphabet, which is finding favour as a medium for Bantu, Negro, and Nilotic languages, consists partly of Roman letters, partly of symbols used in the International Phonetic Alphabet (see below). The Coptic alphabet, still in use in the Coptic Church, and the ancient alphabet of the Goths (the true Gothic alphabet) are mixed alphabets derived mainly from the Greek alphabet. These last are of interest to palæography rather than to contemporary geography, and need no further mention here.

Alphabetical order. As the alphabets of the world developed, the letters settled into a conventional sequence almost universally accepted and never questioned. Why our English alphabet begins with a, b, c, instead of g, w, l, would be hard to explain. Alphabetical order is, nevertheless, one of the fundamental elements of order in Western thought, for upon it depends the whole of our scheme of written arrangements of the names of things. It may be arbitrary, but, whether it be through the dictionary, the telephone directory, or the library catalogue of the British Museum, alphabetical order of reference ensures quick access to all written and printed sources of information. To upset alphabetical order is to cause inconvenience which may be serious. Our English alphabetical order, though rigid as regards the primary sequence of the letters is not so in subordinate sequence. There may be several ways of getting an alphabetical list of headings to run, and that which gives most help to the eye is usually the best. Even the order of some of the letters may not be quite settled for some alphabets (that of ż and ź in Polish, for instance), and the place of a letter in a foreign alphabet may not be that which it occupies in the English alphabet. Vowel-letters which carry diacritical marks in Scandinavian alphabets, for instance, come right at the end.

Foreign alphabets are not necessarily written from left to right like the European alphabets. The Hebrew alphabet, and the Arabic alphabet and its derivatives are written from right to left. In countries which use alphabets written from right to left, pages are usually numbered in the reverse way from ours, and an Arabic, Hebrew or Urdu book therefore begins at what we regard as the end.

Syllabic script. Purely syllabic script has been provided by scholars for some of the American Indian and Eskimo languages and for dialects of the Canadian Arctic, for languages of south-west China, particularly of the Miao group, and elsewhere; but true and exclusively syllabic script is rare in official use.

The script derived from ancient Ethiopic and now in official use in Abyssinia may best be regarded as syllabic, for each distinct symbol represents a consonant combined with a following vowel. In the syllabic table for Amharic, the official language of Abyssinia, the symbols are classed by the vowels represented in them, and are arranged accordingly in seven columns. Allowing for diphthongal combinations and ligatures, the Amharic syllabary is said to include 271 separate symbols.

This Amharic syllabary, which is of ancient use, is such an alien in its surroundings, at least in appearance and structure, as to make the question of its origin one of great interest. In the intense blackness of its symbols and in the regularity of its structure, it bears a superficial resemblance to some Indian scripts (see below). To suppose that it might be of Indian origin is to imply very early relations between India and north-eastern Africa. In view of the part played by India, according to Panikkar, in the containment of Muhammadan naval power in the fifteenth century, it seems at least possible that India had maritime relations with countries to the west long before the Arabian expansion.

A syllabary called the *kana* syllabary is in essential though accessory use for the writing of Japanese. It consists of forty-eight basic symbols, to which twenty-five have been added for the representation of voiced sounds, and a further five for the representation of foreign sounds. There are two kinds of *kana* characters, the *katakana* and the *hiragana*, which differ in function, like our Roman and Italic type. *Kana* is usually written downwards, in columns which range from right to left, but it may be written on lines, usually from left to right though occasionally it will be found to run the other way.

Logographic script. The outstanding modern example of logographic (or ideographic) script is the Chinese, in which thousands of unique characters are used, each representing a whole, simple word. A single character may be formed by as many as seventeen strokes of the brush. The script is some thousands of years old, and many of the characters were, originally, stylized and simplified pictures of the things represented. A given character stands for the same idea throughout China, but the reading, that is, the spoken word which corresponds to the idea, may be different in different parts of China. This means that a speaker of the Cantonese dialect may not even understand the Nanking dialect, but the written language is understood by literate Chinese throughout the country. This literary unity of Chinese is one of the most powerful elements in the political coherence of China.

In spite of their complexity, the Chinese characters can be arranged in regular order. There are several systems of arrangement, the commonest being by **radicals.** The more complex characters consist of two or more parts, one indicating the sound to be attached to the whole, the other indicating the sense, and these parts are in many instances characters in their own right if used alone.

Two hundred and fourteen characters which enter into the composition of other characters have been selected and arranged in the order of the number of separate brush strokes with which they are drawn. In that order, they constitute the headings for the serial arrangement of words into organized lists, and when so used are called radicals. The choice of radicals for arrangement in order, however, is arbitrary; and the radicals themselves, when not used as separate characters, maybe appear in compressed or otherwise modified form.

There are different styles of written characters in Chinese use, which correspond to our Roman, Italic, bold-face, and cursive styles of type. The characters are usually written downwards, in columns which range from right to left, and pages are normally numbered in the reverse way from ours, so that a Chinese book opens at what we regard as the end.

The Chinese have made numerous attempts to simplify the writing of the language, by simplifying the characters and by other means. Chinese is a language composed of a small number of syllables, each of which can be uttered in different ways (and so multiply the number of simple words) and can be combined with other syllables to build compound words. The order of words is very logical and context leaves no doubt as to meaning in the spoken tongue. But the written character ensures the correct identification of things out of context. The syllable *feng*, said in one way, may express one of five different ideas. If written, there would be no doubt about which was meant, each being represented by a unique character. English words, like *bill*, *tear*, *row*, which have different meaning, or both different pronunciation and meaning in different context, have no meaning when they stand alone.

The Chinese character is used in the Chinese way in parts of Indochina, and in a special way in Japan; and it was formerly used in Korea.

Combined systems. The ancient Sanskrit of India and the Pali of Ceylon were the linguistic parents of many scripts now in use in India, Tibet, and Further India, most of which are combinations of alphabet with syllabary. In Sanskrit the distinct symbols for vowels appeared only when they were meant to represent complete syllables and were therefore nearly always at the beginnings of words. The consonantal symbols represented consonantal sounds combined with a following short *a*, and were therefore syllabic symbols. When a symbol was meant to imply a different vowel it was marked with a corresponding vowel-sign, which might be placed before, after, above or below the symbol affected. Numerous compound sounds in Sanskrit were represented by distinct symbols; and numerous symbols which often occurred together came to be represented by ligatures. Ballhorn lists 431 distinct symbols and 174 ligatures.

As the phonetic structure of Sanskrit, like that of many

Indian languages to-day, was more complex than that of most modern European languages, the modern Indian scripts derived from Sanskrit seem, at first acquaintance, to be unduly elaborate, but their complication is necessary. One of the most symmetrical is that used for Tamil, and it can be presented as a table consisting of 12 symbols for vowels and diphthongs, 18 for consonants, and 187 combinations and ligatures for syllables; but a few additional symbols and ligatures have also been found necessary. The most important of the Indian scripts is that used for Hindi. Most of the Indian combined scripts are written from left to right like our own.

A combined script of quite a different sort is in use in Japan, where the native *kana* syllabary has to be used in conjunction with Chinese characters for the adequate writing of the language. The combined script is called *kanamajiri*. Some of the characters used in Japan are no longer in use in China, and a few have been developed by the Japanese themselves. The characters are called *kanji*, and nearly all have both Chinese and Japanese readings, the former derived from early readings in different parts of China. A character may have several Japanese readings. The selection of the correct reading depends upon familiarity with the written language. Like the Chinese, the Japanese have made serious attempts to simplify the forms of characters. The Japanese have even considered simplification of the use of *kana*. Legislation of 1947 limited the number of characters to be used in the press and in official documents to a selected list of 1850.

Korean is written in a script which, compared with those used for Chinese and Japanese, looks deceptively simple. It is most conveniently treated as a combined script, for it has both alphabetical and syllabic characteristics.

CHAPTER IV

TRANSFORMATION OF SYSTEMS

Information about foreign countries comes to us largely in the languages of those countries; and national maps are issued not only in the languages, but in the scripts of the nations which produce them. Greek maps, which are the best possible source of geographical information about Greece, are incomprehensible to those who know no Greek. But geography has now to deal even with official maps in Hebrew and in Sinhalese. Geographical names belonging to all the living languages of the world have to be written on British maps of foreign countries in such a way as to make them easy for English-speaking people to read.

The Roman alphabet is now in official use over a great part of the world, and names written on foreign maps in forms of the Roman alphabet are at least comprehensible to the eye, whether or not we can correctly read the names on them aloud. Names written in scripts other than Roman, however, present the problem of transforming the foreign scripts into corresponding renderings in the Roman alphabet. This can be done by the systematic **transliteration** or transcription of alphabets, and by the **transcription** of syllabic symbols and logographic characters.

The transfer of words from one script into another is superfluous and a mere hindrance to a master of both. No English student of Siamese need write that language in the English alphabet for his own help. But, if he writes about Siam or the Siamese for the English-speaking public; if Siamese news is to be reported in British newspapers; if Siam is to be shown on British maps or in British atlases; if Siamese literature is to be catalogued in British libraries, the need for satisfactory ways of writing Siamese proper names in the English alphabet is immediate. Their transcription is therefore indispensable to public life.

Geographical names have been recorded by travellers, transmitted between nations by mariners, merchants and missionaries, and transcribed from one script into another, from very early times. The problem of how to render them is nevertheless complex, and its solution depends partly upon the kind of name under consideration, but mainly upon the purpose for which transcription is required. The essential difficulty was correctly formulated long ago in *A Survey of the Turkish Empire* (2nd ed., London, 1799), by W. Eton who said: 'Where the orthography of names is established by long custom, we must, I suppose, continue to use it. . . . Where this is not the case, we must spell names as the natives do, if they make use of the Latin characters; if they do not, we must either write their character, or make use of letters of our own alphabet that will produce, as near as possible, the same sounds; or, without regard to the sounds that answer to their letters. I will not decide which method ought to be followed, or which is the real language of a people whose orthography is fixed, the oral or the written. I will only observe, that it is very difficult to render the principal sounds of one language by letters of another, not to mention modifications; and this is particularly so to the English, whose vowels have a very uncertain pronunciation.'

The question is not the simple one of how to write foreign names phonetically in our alphabet, but how, if it be possible, to transcribe them so that they shall preserve their identity well enough to be recognized beyond doubt, either when read aloud, or when compared with the written originals, say on foreign maps. A transcriber from non-alphabetical script will render them sound-for-sound; but a transcriber from alphabetical script has to decide whether to render them sound-for-sound or letter-for-letter, or to compromise between the two methods. Alphabetical script confronts us, from the beginning, with the conflicting claims of the spoken and the written word. Further, since geographical names serve primarily for the certain identification of places, the greatest possible amount of distinction between them is desirable. Distinction is attained by various means, one of the most important of which is inconsistency in the way of spelling or writing similar names. The presence or absence of an accent, of an apostrophe or a hyphen, may be enough to distinguish between two names which would otherwise be the same. It follows that, in the transcription of foreign names, the distinguishing form must be preserved as closely as possible, and the temptation to improve it, for orthographical, grammatical or phonetic reasons, should be resisted.

Transliteration (Latin *litera*, a letter). The transformation of spelling from the letters of one alphabet into the closest corresponding letters of another is called transliteration. Between phonetic alphabets, like the Serb and Croat, the process is nearly, though not quite, one of direct substitution, letter by letter; but there are three ligatures in the Serb alphabet which are represented by digraphs in the Croat. In general, it is rare to find that any two alphabets, particularly if they be old or long established, are corresponding sets of equivalent signs. They may not even have the same number of letters, Russian now having thirty-three to our twenty-six. It formerly had thirty-seven. The question of the transliteration of Russian is, indeed, an admirable example of the complexity of the problem of transliteration.

By Russian we mean Great Russian, the language of Muscovy, which is now accepted as the official language of the Union of Soviet Socialist Republics. The oldest known Russian documents are in Cyrillic script, but very early documents in Glagolitic or 'spectacle' script are also known. The Cyrillic is an alphabetical script derived from Greek liturgical uncial letters of the ninth century, given, many believe, to the Orthodox Slavs by St. Cyril (827–869 A.D.). There is reason to suppose that Cyrillic and Glagolitic were inventions of the same mind, the former intended for Greek Orthodox, the latter for Roman Catholic Slavonic use. The Cyrillic became the accepted alphabet for the languages called Russian, and for Bulgarian and Serbian. The Glagolitic survives in Dalmatia, but only in liturgical use. Once established, the Cyrillic alphabet underwent little change in Russia until the time of Peter the Great (1672–1725). Amongst his many reforms was that of Russian printing, and in 1708 he ordained that all secular books were to be printed in 'civil' type. The Old Slavonic forms of the letters remained in liturgical use. In 'civil' type the Russian Cyrillic letters assumed their modern forms, and the Russian alphabet underwent no further change until 1917, when i, ѣ, ѳ and *ѵ* were abolished and the use of ъ was limited to that of a stop after certain prefixes.

The Russian alphabet is well adapted to the phonetic requirements of Russian. It has a letter for every Russian consonantal sound, it distinguishes between so-called 'hard' and 'soft' vowels and, to a certain degree, can render palatalization (the imparting of a liquid sound to some consonants by momentarily touching the roof of the mouth with the flat of the tongue). Writing is therefore fairly compact. Modern Russian, nevertheless, is by no means phonetically written; some Russian sounds, e.g., those of ы and hard л (the second *l* in little) have no counterparts in most Western European languages; and others can be represented only by clumsy combinations, e.g. *schtsch*, which is German for щ. The transliteration, and even the transcription, of Russian into Western European alphabets is therefore difficult. The problem has been approached in the following ways:

(1) *From outside*

(*a*) As part of the general problem of the romanization (that is, of the rendering in Roman letters) of all Slavonic languages written in Cyrillic script. For Russian can be written in (not merely transliterated into) the Roman alphabets which have been developed for other Slavonic languages such as Czech and Polish. The Czech alphabet, invented by John Huss (1373–1415) was further developed for the writing of Croat and Slovene by Vuk Karadžić (1787–1864). In an amplified form the Croat alphabet is still strictly phonetic and capable of such wide application that it has found favour not only amongst Slavonic scholars, but also amongst other linguists, and has been used even in scientific grammars of Oriental languages. It has provided the basis for a system of romanization of all languages written in Cyrillic script which is already in limited international use. That is to say, a satisfactory international romanization of Russian has already been achieved, from outside. Damiani calls it the 'scientific system', and Brauner strongly advocated its auxiliary use in Great Britain.

(*b*) As a problem for nations which need to render Russian into scripts which they can read. Eton, who had spoken Russian for twenty-three years, chose to write Russian proper names in their Polish forms rather than transliterate them into the English alphabet, but was only offered instruction, by the reviewers of his time, for his pains. The Austro-Hungarian General Staff, in the early nineties of last century, followed the same indirect procedure, by accepting the romanization of Russian geographical names through the Polish alphabet, thus avoiding the use of two systems of Roman orthography on its maps of the northern frontier of the Empire. In general, however, the problem has been faced directly, and most Western nations have their own systems for the rendering of Russian names on maps.

(2) *From inside*

(*a*) As a Russian problem of how to produce a system of romanization which might be accepted abroad. In 1907 the Imperial Academy of Sciences of St. Petersburg published particulars of a system, but it met with little response outside Russia. In 1935 the All-Union Committee of Standardization of the Russian Council of Labour and Defence announced a system ('All-Union Standard 8483') in which no diacritical marks were used. It has not been accorded full recognition in Russia and has not been applied abroad.

(*b*) As an unofficial Russian problem of how to render Russian in particular Roman alphabets. In 1934 the *Moscow Daily News* published a remarkably good though not wholly satisfactory scheme for the transliteration of Russian into English letters. As the work of Russian-speaking English and American journalists in daily association with educated Russians, it was produced in favourable circumstances. It marks an important stage in the development of a system which will enable English people to pronounce Russian names intelligibly.

So great was the Russian contribution to science and art during the nineteenth century that the diversity of national forms in which the names of Russian scientists, artists, and statesmen are known, is an accumulation of intellectual rubbish. Пушкин has become Pushkin (English), Pouchkine (French), Puschkin (German), Pusckin (Italian), Puszkin (Polish), Puşkin (Romanian), Poeshkin (Dutch); and, in this favoured example, the initial letter is not affected. It would be wise for all, Russians included, to accept the scientific transliteration 'Puškin' when precise rendering is required, and to discard the rest.

This, however, is not the only difficulty. There are at least ten different systems for the transliteration of Russian in use in Great Britain and the United States. Some, like that of the Library of Congress, are systems of rigorous transliteration; others, like that of the *Moscow Daily News* approach phonetic transcription. They differ little in the rendering of consonants, but exhibit various ways of rendering й particularly in adjectival endings, and of ё and ы. The main difficulty of transcribing Russian into English letters (transcribing implies some attempt to render pronunciation) is that in Russian yodization (the emergence of the y-sound before and between certain vowels) and palatalization vary in degree.

It should be clear that successful transliteration of a foreign alphabet for a given purpose depends greatly upon the choice of the system used. The names rendered from foreign alphabets on British official maps and charts are transliterated in accordance with the recommendations of *Alphabets of Foreign Languages* by Lord Edward Gleichen and John H. Reynolds (2nd ed., reprinted, London, 1951). Extensive changes in orthography abroad demand the preparation of a new edition of this work, but the labour involved in collecting the information and presenting it in a way appropriate to our times will take some years. The new edition is therefore to appear in fasicules.

Those new to the procedure are apt to expect transliteration to produce perfect results (that is, to render both foreign spelling and foreign pronunciation); and they expect to find transliteration-tables which can be applied by rule-of-thumb by persons who know nothing of the languages and alphabets concerned; moreover, they expect the process to be reversible.

Transliteration will so work only between two alphabets

which are phonetic, and which represent the same sounds. Between the Serb and the Croat alphabets the correspondence is almost complete. Between the old and the new Turkish alphabets there is little direct correspondence at all, and, although the old can easily be transliterated into the new, the reverse is possible only to persons with an intimate knowledge of old Turkish orthography. If a system is reversible it is so by chance rather than by design. Even with good systems, transliteration back into the original language can sometimes produce bizarre results. The mineral-name gaylussite has been fed into the Russian mill to emerge as halloysite, that of quite a different mineral, on reverse.

In written Arabic and Persian only the long vowels are commonly represented. On transliteration, the letters for short vowels have to be supplied, and that can be done only by persons conversant with the languages.

The most informing statements I have read on the problems of transliteration are the contributions of King Vajiravudh of Siam to the official discussion of the transliteration of Siamese.[1] The king was an excellent linguist, a fluent writer of English, a witty debater, and a good geographer, and his arguments and examples make entertaining reading on difficult subject matter. What he had to say over forty years ago can still be read with profit by anybody interested in transliteration.

Those scripts (principally Indian) which represent a combination of alphabet and syllabary, can mostly be rendered into Roman letters symbol-for-symbol rather than sound-for-sound. Names in the languages of India and Ceylon, however, were written on maps in an entirely different way during the period of British rule (see below). During that period India was responsible for the mapping

[1] *J. Siam Soc.*, 9, pt. 4 (1912), pp. 1–10; and 10, pt. 4 (1913), pp. 25–33.

of certain neighbouring and Middle Eastern countries, and British official transcription of Tibetan names was according to Indian departmental rules.[1] Excellent official tables were devised for application to Burmese [2] and Shan names.[3]

Transcription (Latin *scriptum*, written). The ideal way of transcribing geographical names from purely syllabic and logographic script would be to render them sound-for-sound in an international Roman alphabet. But no such alphabet exists, and the resources of the European alphabets do not permit of the close rendering of many Oriental sounds. Moreover, as the meaning of a syllable depends, in some languages, upon the tone of voice in which it is uttered, and in different languages there are different numbers of significant tones, the representation of tone is necessary for precision, but it introduces typographical complications little suited for use on maps. The only syllabary discussed by Gleichen and Reynolds (and that incompletely) is the Amharic. Rather than develop British systems for the transcription of syllabic script, our official mapping and charting authorities have preferred to accept standard systems, such as the Wade-Giles system for Chinese [4] and the Hepburn system for Japanese.[5]

Both transliteration and transcription present two cases for consideration: that of names in the official language, and that of names in subordinate languages written in the official script. In Greece there are still many Turkish and

[1] O'Connor, W. F., and C. A. Bell. Rules for the phonetic transcription into English of Tibetan words. India, *Foreign Dept. Proc.*, June, 1904.

[2] Tables for the transliteration of Burmese into English. 3rd ed., *Rangoon*, 1908.

[3] Tables for the transliteration of Shan names into English. *Rangoon*, 1892.

[4] Giles, H. A. A Chinese–English Dictionary, 2nd ed., 3 vols., *London*, 1909–1912. Ref. in vol. i, pp. xvii–xviii.

[5] Hepburn, J. C. A Japanese–English and English - Japanese Dictionary. 4th ed., Tokyo, 1888, p. iii.

Bulgarian place-names written in the Greek alphabet on Greek maps. If transliterated as Greek the results may be very clumsy, but if the Greek conventions for rendering Turkish and Bulgarian sounds be taken into consideration, desirable simplification is possible. No general way of dealing with this difficulty has yet been devised.

The most comprehensive statement on transliteration and transcription which I have encountered in English is Wharton's essay in the *Transactions of the Philological Society*.[1] It is one of the fundamental documents relating to the transference of words from the symbols of one script into those of another.

In established systems of transcription and transliteration there are very strong vested interests, for, once a system has been brought into use in a big library, or applied to extensive series of maps or charts, it cannot be changed without great expense, delay, and confusion. It is therefore most important that new systems be applied experimentally to trial documents before they are announced. Had the P.C.G.N. system for Greek been applied experimentally to one sheet of a Greek official map of Greece before the publication of the first edition of *Alphabets of Foreign Languages* (1921), serious delay of work which took place in 1941 might have been avoided; but the system as announced in 1921 lay untried until 1941, when it was found to be quite unsuited to modern Greek.

The transliteration or transcription of names on foreign maps and charts is best done on **name-lists** or on **name-traces.** In either case, the sheet or chart under consideration should be lightly 'squared up' in pencil; that is, the surface should be divided into quadrilaterals of one minute of longitude and one of latitude. This can usually be done by ruling lines from corresponding points on opposite,

[1] 1921–1924 (1928), pp. 59–112. Select bibliography in chronological order.

graduated margins. The squaring serves for the recording of the positions of places to the nearest minute.

The transliterated names may be listed in the order of the run of the minute quadrilaterals in which they begin, read row by row and downwards, like the lines of a printed page; or they may be written directly on tracing cloth or tracing paper placed over the sheet. The name-trace should bear guide lines, to ensure its accurate replacement over the sheet after removal.

Conversion of systems. It has sometimes been found expedient to reduce certain foreign spellings of names to supposedly English form by substituting English letters or combinations of letters for those used in foreign ways. For instance, in copying French maps of parts of Africa British authorities have at times seen fit to change the French *djebel* into *jebel*, *dj* in French being the equivalent of *j* in English. In the same way the Italian Scioa becomes Showa.

Conversion tables for the change of French, German, and Portuguese spellings into English form exist, and have at times been extensively applied. The procedure is most questionable and should not be followed if other measures are possible, for it 'cannot be trusted to produce the right result automatically'. A remarkable example of the wrong kind of result is given on p. 229 of the *Geographical Journal*, 52 (1918).

Fortunately, many written foreign geographical names can be satisfactorily rendered into English form for cartographic purposes without mastery of the languages to which they belong; but, whoever undertakes their reduction from foreign original sources, will find it necessary to understand the manner in which the names are used in the countries concerned, the phonetic structure of the languages under consideration, and the grammar of their articles, nouns, and adjectives.

CHAPTER V

THE IDENTIFICATION OF PLACES

Places are identified by name and **position.** If the position be required of a place known by name, but known merely to be in a particular country, the map is the most inconvenient document in which to seek it; for the map may be large, or in many sheets, each of which has to be read in a systematic way, say by quadrilaterals of the mesh of latitude and longitude, row by row from top left to bottom right, until one finds the place. It is much more convenient to seek it first in an alphabetical list of names.

Such lists exist, in travel time-tables, in the indexes to serial geographical publications, to official manuals and reports on countries, to books on geographical regions, to standard atlases and maps, and to large-scale maps published as series of sheets, and in postal guides, census reports, and **gazetteers.** Sources like these are so valuable as to deserve the special attention of cataloguers in geographical libraries.

Most indexes to maps, and to sets of sheets, consisting as they do of the names appearing on the sheets, are uncritical documents. They are, as a rule, official productions which were prepared *pari passu* with the sheets to which they apply, and theiı preparation involved no criticism of the state of the names on the sheets.

Numerous indexes to series of sheets and charts were issued by different Allied authorities during the wars of 1914–1918 and 1939–1945, some of which are now available in geographical libraries. They are very useful, though not authoritative because few if any of them were prepared by the native governments of the areas represented on the sheets.

In the eighteenth century a journalist was called a gazetteer, and certain compendia of geographical information prepared for the use of journalists were then called 'the gazetteer's or newsman's guide', or simply, 'the gazetteer'. Such compendia were in official use in China long before they were thought of in Europe, and the ancient Chinese provincial gazetteers are well known to sinologists. During the nineteenth century it was the duty of certain officials in British possessions, and of certain British representatives in other countries, to maintain gazetteers of areas for which they were responsible to H.M. Government. The district gazetteers produced in India are a lasting record of conscientious activity.

A modern gazetteer is a geographical dictionary, giving full and authoritative information about all the places and features listed by name in it. The research which goes into the production of a true gazetteer is so thorough that, given stability of conditions, much of the information may remain valid for a long time. Countries in which geographical names and administrative boundaries undergo little change may therefore be able to keep their gazetteers up to date with little trouble. Bartholomew's *Survey Gazetteer of the British Isles*, for instance, is now in its ninth edition (Edinburgh, 1943). On account of the difficulties of revising large and exhaustive works, few foreign national gazetteers are really up to date. Some of the larger productions of the past are nevertheless extremely useful to informed persons. The following deserve particular mention: Madoz's *Diccionario geográfico-estadístico-histórico de España y sus posesiones de ultramar* (16 vols., Madrid, 1846–1850), *Słownik Geograficzny Królestwa Polskiego* (14 vols., Warszawa, 1880–1895), Joanne's *Dictionnaire géographique et administratif de la France et de ses Colonies* (7 vols. and introduction, Paris, 1890–1905), *Norges Land og Folk* (29 vols., Christiania, 1898-Kristiania, 1915), *The*

Imperial Gazetteer of India (new ed., 25 vols. and atlas, Oxford, 1907–1909 and 1931), and Sawada's 日本地名大辭典 (*Dictionary of Japanese Geographical Names*, 6 vols., Tokyo, 1937–1938).

The political adjustments and readjustments which have been taking place all over the world since 1914, have virtually suspended the production of new national gazetteers in many volumes; but useful gazetteers of departments, provinces, and states are now being issued in both North and South America.[1] To undertake the preparation of an exhaustive national gazetteer would amount now, on the part of a country concerned, to a declaration of faith in a forthcoming period of prolonged peace.

Up to 1914 there were several world-gazetteers which used to appear in new editions from time to time, for instance, Ritter's *Geographisches Lexikon*, of 1904–1906 and Lippincott's successor of 1905 and 1912 to Longman's gazetteer of 1895. But, except for the *Dictionnaire des Bureaux de Poste* (5th ed., 2 vols., Berne, 1951), which is a complete and indispensable list of all the post offices in the world, no exhaustive world-gazetteer appeared after 1912 until *The Columbia Lippincott Gazetteer of the World* was published in 1952. A world-gazetteer is not a substitute, country by country, for the national gazetteers which the world needs; but conversely, there is now nothing better available for many countries than the Columbia Lippincott. One of the most important of world-gazetteers has no separate existence. It consists of the indexes to the seventy odd volumes of the Admiralty *Pilot* series, also known as the Admiralty *Sailing Directions*, which are under constant revision, and to which cumulative supplements are regularly

[1] *See* Gazetteers of American states and territories, *Geogr. J.*, 106 (1945), pp. 162–163, *and* Work on South American place-names, *Ib.*, 109 (1947), pp. 151–153.

issued. The indexes to these volumes and to their supplements constitute an authoritative and up-to-date gazetteer to all parts of the world accessible to British ships.

A place may be found on a map or chart if particulars of its position be given according to some system of coördinates, or according to the distance and direction in which it lies from some other place. The other place need not be a centre of population, and may be a critical point such as the mouth of a river, or confluence of two rivers, or the point where a canal crosses a frontier. The system of localization chosen for a particular list of names will depend upon the character of the materials from which it was compiled or on the quality of the survey of the country or countries concerned, or on both. For countries like Afghanistan, of which the maps are not derived from accurate modern surveys, places are best localized by direction and distance from other places. But for western European countries, of which the maps are based upon rigorous geodetic survey, coördinates (such as latitude and longitude) may prove to be best. But they may be of several kinds.

A map or chart is the result of extremely complex measurements and calculations whereby a part of the Earth's *curved* surface is represented on *flat* paper. Anything drawn on so large a scale that the difference between the curvature of the earth's surface and the flatness of the paper does not matter, is not a map but a plan. A few of the more important ways of projecting part or whole of the round surface of a sphere on to a flat surface are diagrammatically explained in the introductory notes to any good atlas. It should suffice here to say that no method of **projection** is ideal, that all produce distortion, but that certain of them have particular advantages over others. When a map is to be drawn, a method of projection suitable for the purpose which the map is to serve must be selected, the projection of the meridians and parallels be drawn accordingly, and the outlines of the maps then drawn in relation to them.

Most British maps are graduated in degrees and minutes of latitude from the Equator, and of longitude measured from the Greenwich meridian. Foreign maps, however, are not necessarily so graduated. Certain French maps, particularly those of Madagascar, are graduated in grads, that is, in hundredths of the quadrant, not ninetieths, as with degrees. Nor are French maps necessarily based on the Greenwich meridian. They are more likely to be based on that of Paris, which is 2° 20′ 14″ East of Greenwich. German maps may be based on Potsdam or on Ferro (in the Azores, and Ptolemy's zero meridian), Austrian maps on Ferro; and the German and Austrian values for Ferro are not the same.[1] Swedish maps may be on Stockholm, Russian on Pulkovo, Greek on Athens, Italian on one of several observatories near Rome, and so on.

The zero meridian used in a particular country for the statement of longitude is called its **prime meridian,** and the difference between any foreign prime meridian and that of Greenwich must be known if we are to reduce the values of the positions on many foreign maps accurately to those of Greenwich. We need to note also, of places shown on maps based on meridians other than Greenwich, whether they be to the east or west of the relevant prime meridian. Grads, referring as they do to systems of meridians of longitude and parallels of latitude, may be regarded as a French version of latitude and longitude.

On some good maps (de Agostini's map of Albania on the scale of 1:300,000, issued in 1939, for instance) no meridians or parallels may appear. Places can be localized on it only by giving their direction and distance from other

[1] Hinks, A. R. Longitudes from Ferro. *Geogr. J.*, 46 (1915), pp. 47–48.

places. If we indicate that a place is 10 kilometres north-north-west of Korçe, others can mark its position on this map.

Meridians and parallels need not appear on a map when it is ultimately printed. For certain purposes (mainly military) it is considered advantageous to replace them by a network of lines at right angles, which divide the map up into regular series of squares or rectangles, which can be numbered in various ways. The network is called a **grid,** and any point on the map can be indicated by rectangular coördinates stated according to the scheme of graduation chosen for the particular grid. The grids on certain maps of Great Britain are graduated from a point to the south-west of Cornwall called the origin. A grid, being an arbitrary device, can be printed on any map and in any way desired; but in practice a grid is superimposed on a map with great care, so that its relation to the original projection is as simple as possible, and the coördinates are easy to read off. A grid cannot, however, be extended indefinitely from its origin, because, being a rectangular system of lines, the meridians and parallels on which the map was really constructed would, were they also shown, 'run away' further and further from the grid lines the further the latter were extended. 'Grid north' bears a more divergent relation to true north and south lines on the map, the further east or west the grid extends from its origin.

In seeking to identify a place known only by name, the finding of the name in a trustworthy index, list or gazetteer is not necessarily the end of all difficulty. One may turn to the atlas, to find nothing marked at the position indicated in the list. It may be necessary, when dealing with obscure places and minor features, to examine the large-scale maps of the country concerned, only to find that the names are given in a foreign script; or that the position given in the list refers to Greenwich while the map is graduated from a different meridian. One may have to decide how the name would be transcribed from the Roman back into the original script; or may have to convert the longitude to that of the map under consideration.

It is well to remember that some gazetteers, like the P.C.G.N. gazetteers of Greece and Albania, though gazetteers of countries, are not national gazetteers. They are, in these instances, British gazetteers of the countries named. Prepared as guides to foreign maps of Greece and Albania, they give positions in terms of the foreign maps, and no useful purpose would have been served by reducing the values to Greenwich.

CHAPTER VI

ENGLISH USAGE

That established English geographical names are the products of English usage is implicit in the consideration I have given to these names above (pp. 17–27). It is time now to consider more closely some of our habits in the use of geographical names.

The giving of names. The spontaneous art of naming things, as revealed in the Book of Genesis, seems to have been lost by all but primitive peoples, the Australian aborigines being amongst the few who still seem to have no difficulty in inventing new names for familiar things in new circumstances, in accordance with the requirements of *tabu*. It would be hard to say when a really new proper name had to be found for any place in England. Letchworth, Welwyn, Immingham, Crawley, etc., though newlydeveloped places, have old names. And it might be harder still to say how most of the innumerable places in England were given the names they bear, except in the very general way of recognizing that many place-names are dedications, designations, and descriptions.

The giving of new geographical names is now an activity practically confined to explorers in the polar wastes, oceanographers investigating the depths, and mountaineers making known the topography of extreme heights. In the nineteenth century during the later phases of exploration by land English explorers were guided by the excellent recommendations of the Royal Geographical Society, which restrained them from commemorating themselves, and from giving commonplace, frivolous or otherwise objectionable names to unnamed features. They are now under further restraint, through the diminution of the extent of unclaimed territory, and new names require the approval of the government of the territory in which it is proposed to apply them. Naming is, in fact, a troublesome problem for polar explorers, for there are now too many rocks called 'the Pulpit' or 'the Spire', and such names as 'Detroit Aviation Society Plateau' are rightly considered to be clumsy. Recent work in the Antarctic suggests that the giving of names might profitably be left to the few more imaginative members of the expeditions concerned.

Authority for names. Some names are right, others wrong. There is a notable amount of legislation with respect to the names of persons and corporations, for they are identified primarily by name.[1] Geographical names are rarely open to dispute at law, though I once waited all day in Court on the chance of being called to make a discouraging statement on the names Near, Middle, and Far East.[2] Occasionally the terms of an insurance policy may raise the question 'Is Cyprus—or Iceland—part of Europe?' To such questions ('Is Monmouthshire part of Wales?' and 'What is the capital of Kent?') there is no clear and final answer, but the need to ask them could be eliminated in many instances by more careful drafting of terms.

The correct forms of toponyms in civilized countries can generally be established from the archives of local government, and those of the administrative subdivisions of such countries can be established from the enactments, including treaty enactments, of the supreme governments over areas in which names may be in question. In Scotland there is no doubt between the names Dumbarton and Dunbarton: all historical documents are said to use the first for the city, the second for the county. What is wrong is to confuse them. In both England and Scotland there is doubt as to the

[1] *See* Name, Encyclop. Brit., 11th and 14th eds.

[2] See *The Times*, 28th, 30th June, 2nd July, 1945 ; and *Hansard*, 16th April, 1946, pp. 2523–2524.

claims of the names 'Britain' and 'Great Britain', which led to an interesting correspondence in *The Times*, conducted partly by the Royal Society of St. George and the St. Andrew Society.[1] The name 'British Commonwealth' came under question at the same time.[2]

In England the correct name of a community is considered to be that used by the community itself; and the correct name of any place or geographical entity in the country is considered to be that used by those most familiar with the place. In so old a country nearly every place-name (toponym) has a long documentary history (in land-title deeds, cadastral plans, and administrative enactments) by which local usage can in most instances be checked. Such a check is in course of application through the Survey of English Place-Names.

The entry of names on the charts of our coasts is undertaken by the Admiralty, and on our maps of the country by the Ordnance Survey, which makes these Departments not the authorities for the names but the authorities merely for recording them. Local opinion may here and there disagree with the work (usually excellent) of the Admiralty or Ordnance Survey; and in rare instances the Home Office, the Post Office, British Railways, or the local government authorities may use a name different from that shown on the map or chart. We may therefore say that, for most places in England, there are generally received names in regular use, and a few for which there are Departmental (not national) official variants. In practice, if we want to find out where a place is, we have to identify it from the name shown on the map or chart; but if its name be in doubt, neither the map nor the chart is necessarily final evidence of the truth.

[1] 21st, 24th, 29th Sept.; 4th 11th Oct.; 29th Dec., 1948; *and* 1st, 3rd, 5th, 13th and 17th, Jan., 1949.
[2] *The Times*, 14th Jan., 1949.

In general, the use of names for polities, communities, bounded areas, and *lieux-dits* rests on some form of legal title. This is perhaps why our geographical nomenclature is both conservative and stable. Change is effected by local agitation and is not imposed by Government. In February, 1955, the Rural Council of Braintree, Essex, corrected the Essex County Council on the spelling of the names Faulkbourne and Hatfield Peveril.

Authority for English geographical names (exonyms) is neither so easily nor so clearly established as for the place-names (toponyms) of England; and the same might be said for place-names in any other country and the geographical names in its language. It is necessary to establish how long, by what authorities, and in what forms, a name has been used in the language concerned, and to do so it may be necessary to examine many documents and maps. Accra, for instance, is a name which means little to natives of the Gold Coast, and none of the local peoples, Twi and Gã, would ever use it when speaking their own languages. They would call it Ngkrang and Gã, respectively. Accra is a European corruption, probably of the form Ngkrang, a mere 'map name'; but it happens to be the British official name of the place, and has been established in English, and used on English maps and charts, ever since Great Britain acquired her rights in the Gold Coast. It is indisputably the English name of the place. This does not mean that we should not ascertain and recognize the several native names; but the place is so well known to us that we are entitled to call it by the name which has become established in English.

The name Kangchenjunga has recently been questioned, because the Survey of India, since 1934, has been using the form Kanchenjunga. Sanskrit origin was claimed for the name in 1930, but, when challenged, the Sanskrit scholars were unable to produce any evidence to support the claim.

The Tibetan scholars, on the other hand, brought proof, from Tibetan literature, that it is of Tibetan origin (as the Maharajah of Sikkim had maintained) the Tibetan form, phonetically transcribed, being Kang-chen-dzö-nga. The form Kangchenjunga has been established in English ever since 1903, and there is no need to displace it from the language.[1] A similar storm of discussion once raged over the English geographical name Karakoram.[2]

The writing of names. While the Chartered Companies were deepening our commercial interest in places overseas, exploration began to give way more and more to survey. The East India Company, for instance, had its own observatory at Madras. The increasing knowledge of far countries brought the need for precision in the writing of the names of far places, and English interest in the writing of foreign geographical names became conscious, so to speak, while Dr. Samuel Johnson was working on his dictionary.

Johnson issued his 'plan' in 1747 and in that very year an anonymous editor (now known to have been Bradock Mead, *alias* John Green) saw his own work completed in Thomas Astley's *New General Collection of Voyages and Travels* (4 vols., London, 1745–1747). Mead used a conventional English alphabet and a systematically romanized Chinese syllabary for the spelling of the geographical names in this work, and seems to have been the first to have used the digraphs *dh* and *zh* for the sounds of *th* in this and *s* in pleasure respectively. His spellings, however, were not widely accepted. In 1787 Gilchrist's dictionary of 'Hindoostanee' began to appear, in which

[1] *See* Burrard and Hayden, A Sketch of the Geography and Geology of the Himalaya Mountains and Tibet, Pt. I, The High Peaks of Asia, 2nd ed., *Delhi*, 1933, pp. 25–33 and 207–212; also *Himalayan J.*, 1931, pp. 152–155, and 1932, pp. 198–214.

[2] *Empire Surv. Rev.*, 3 (1935–1936), pp. 115–120.

Hindi and Urdu words assumed clumsy romanized forms. In 1788 Sir William Jones published his dissertation 'On the orthography of Asiatick words in Roman letters', in which he recommended the principle of using our symbols for consonants as in English but those for vowels as in Italian. This was a sound phonetic principle and was accepted by many Orientalists, but it was long before anything so satisfactory was generally applied. Cook, so far as I know, had no instructions about the writing of native names; and as late as 1799 we find the editor of the account of the voyage of the ship *Duff* trying, in vain, to regulate the spelling of names used by the South Sea Islanders.

In 1836 an enlightened naval officer, Commander (later Rear-Admiral) John Washington, became Secretary of the Royal Geographical Society, and in that year the Society tacitly accepted Sir William Jones's principle and applied it to the writing of foreign geographical names. In 1841 Washington returned to the Admiralty, where he developed a conventional alphabet on Jones's principle and applied it in vocabularies of the Eskimo dialects of Labrador (1850) and Greenland (1853). In 1862 the Admiralty issued important instructions to hydrographic surveyors which embodied recommendations on the handling of geographical names, and in 1884 it issued rules for spelling them. The Royal Geographical Society adopted the same rules in 1885, whereafter they became known as the 'R.G.S. System'. The system was quickly accepted by the War Office, Foreign Office, and Colonial Office.

It was developed while grand exploration on land, and colonial expansion, were in progress. Many native languages were then being reduced to writing for the first time, mainly by missionaries; very few people indeed were proficient in Colonial languages, particularly in Africa, and it was widely supposed that African languages were primitive and imperfect media of communication. It was necessary to have a consistent British way of spelling the geographical names in colonies, and the R.G.S. system provided means adapted to the temper of the time. Thousands of native names were recorded in the R.G.S. alphabet and entered on British maps of various parts of the world; but it is important to note that, so far as I am aware, not a single native language has ever been reduced to writing in this system.

Immediately after the war of 1914–1918 the Admiralty considered the possibility of adopting a 'universal' system of geographical nomenclature, and, with the indispensable assistance of the Royal Geographical Society, formed an inter-Departmental Committee, the Permanent Committee on Geographical Names for British Official Use (P.C.G.N.) to advise on questions of principle. The P.C.G.N. undertook the revision of the R.G.S. System and produced the R.G.S. II System, formulated rules for spelling and issued transliteration tables (1921), published exemplary lists of names in the succession states to the German, Austro-Hungarian, and Turkish Empires, and came to be regarded abroad as what it is not: the British statutory authority on geographical names.

Since 1918 the world has undergone fundamental geographical changes. Nearly every part of the land-surface of the globe is now under the control of sovereign power. Major exploration is over, except in the Antarctic, and the intensive study of ocean-depths and great mountain-heights is taking its place. Nearly the whole world has been mapped, though with varying degrees of precision in different regions, and the surveying, whether geodetic, topographical or hydrographic, is being done by the sovereign governments or their colonial administrations. Certain countries which have recently become independent, like Indonesia and Libya, are presumably still using the results of survey by Imperial authorities since withdrawn, but will ultimately establish their own mapping and charting services.

The foreign official map or chart is now our readiest source of geographical information about other countries, and never again will British explorers or surveyors have easy access to some parts of the world of which they produced the first reliable maps and charts. Native languages, too, are being more and more closely studied by anthropologists, linguists, and phoneticians, and their orthography is becoming the concern of governments. There is less recording of geographical names never before reduced to writing, but an urgent need for the adequate transliteration and transcription of those now being written in new or reformed alphabets and scripts.

The R.G.S. System, as its originators were well aware, was a conventional system and no more phonetic than most other alphabetical systems. But it has been 'taken (in flat contradiction of the declaration of its authors) to be a phonetic system' (Wharton). Its originators took 'the true sound of the word as locally pronounced' for the basis of spelling, and aimed at an 'approximation . . . to the sound. A system which would attempt to represent the more delicate inflexions of sound and accent would be so complicated as to defeat itself' (Rules of the R.G.S. System as given in 1885). The originators of the broader R.G.S. II System announced their alphabet as phonetic (preface to *Alphabets of Foreign Languages*, 1921, and more specifically in that to the second edition of 1933). In their *Rules for Spelling* they also repeated the statement that 'every letter is pronounced', but eight consonantal sounds were represented in their alphabet by digraphs, of which the letters were not separately pronounced; and they said that 'The system aims at giving a close approximation to the local pronunciation'. By so much had the aural perception of English geographers become debased between 1885 and 1921! Nobody had ever noticed the difference, and I was rebuked by the late Mr. A. R. Hinks for pointing it out in 1942. The measure of recovery in recent years can be gauged by comparing the P.C.G.N. review,[1] of 1930, of proposals on the practical orthography of African languages put forward by the International Institute of African Languages and Cultures, with the Committee's own proposals of 1950–53 for the spelling of African place-names.[2]

The P.C.G.N. having carefully considered conditions in the world to-day, has recently (1954) transformed the R.G.S. II System into *Principles of Geographical Nomenclature* and has developed a conventional English alphabet for the transcription and transliteration of geographical names better adapted to the needs of our time than that used in the earlier systems.

The recording of names. In the earliest edition (1849) of *A Manual of Scientific Enquiry* the Admiralty published instructions for the collection and recording of foreign geographical names by its surveyors. The R.G.S. system eventually became the basis of rules for the collection and recording of native names issued by Colonial Governments, and the methods developed are well illustrated in an account of work by an experienced officer in a difficult area, recently published in the *Geographical Journal.*[3] In his book on *The Phonetics of the Hottentot Language* (Cambridge, 1938) D. M. Beach has shown how rough and corrupt the amateur recording of many native languages must be. As Professor Sir Hamilton Gibb once said to me, 'People do not hear sounds which they do

[1] *Geogr. J.*, 71 (1928), p. 296.

[2] *P.C.G.N. Leaflets, No. 6*, Oct., 1950; *6a*, July, 1952; *addendum*, Nov., 1953.

[3] 109 (1947), pp. 99–102.

not know', and Beach states that 'Without a great deal of training the hearing of most persons is very inaccurate and limited in scope', remarks which should be taken to heart by all who attempt to record native names.

The older methods used in the field were not satisfactory. They were applied as a rule by surveyors, and ability in surveying is not necessarily accompanied by ability to hear the sounds of strange languages. The informants chosen might not be suitable, and numerous badly and wrongly recorded names represent misunderstanding between informant and collector. Most persons familiar with the procedure are aware that some supposed native names on colonial maps mean 'over there', 'I don't know', and similar uncomprehending answers to the question 'What do you call that hill?' The endeavour now is to get native names recorded preferably by phoneticians, or by anthropologists with linguistic training, and to record them either mechanically, or by writing them down in International Phonetic symbols or in the vernacular alphabet if there be a suitable one in use. And when recorded they must be checked by being read back to native critics.

Correction of names. Our attitude towards geographical names in other languages is shown by our treatment of Welsh and Irish names. We are content to use anglicized forms of the names of well-known places and geographical features in Wales and Ireland, leaving it to the Welsh and the Irish themselves to produce maps of their countries wholly in their own languages, should they need them. We have traditionally used English, anglicized, and conventionally English names for those parts, features, and places of the outside world most familiar to us. We have resisted attempts to introduce so-called better spelling of well-known English names of foreign places, like that of Astley in the middle of the eighteenth century, who tried to introduce Kairo, Mekka, Kongo, etc. And, unlike the

Germans, we accept familiar foreign names without regard to their meaning. Such a name as Nizhni Novgorod (now Gorki) had no more meaning for us than the name Thomas Atkins, and to be told that the former merely meant 'lower new town' would be enough to dissuade us from translating foreign names which have meaning. We manage, too, to use many foreign names without the grammatical barbarism of adding English topographical terms which duplicate what is already said in the names (pleonasm). It is now rare to hear reference to 'the Hoang *Ho river*', to '*Mount* Fuji *San*', or '*Lake* Victoria *Nyanza*'.

We are easy-going with regard to our geographical names, content, in the main, to accept them as they are given to us by those who produce the maps. But some names are right, others wrong. It was necessary, in the course of time, to correct Cook's Otahiti (for Tahiti) and Bolabola (for Porapora). Revision and correction of names is therefore an indispensable activity in cartography. But it is one fraught with great temptations, especially to those who have, or think they have, the requisite linguistic knowledge. Old Admiralty instructions (of 1862) to hydrographic surveyors tell the surveyor that 'He should beware of meddling with names already established' and the warning is printed in italics.

The correction of geographical names is more than a linguistic or orthographical matter. The main function of geographical names is **to ensure the certain identification of places,** and not necessarily to preserve native words like bees in amber. If a name becomes established as the name of a place in a particular language, it matters little whether or not it be linguistically pure or rationally written. This is well illustrated by Dr. H. G. Fourcade's consideration of the name Zitzikamma [1] which, after 130 years of stability of

[1] *South African Surv. J.*, V, pt. 4, Jan., 1939, pp. 165–169.

form, was considered for improvement. 'Apparently' wrote Dr. Fourcade, 'the intention . . . is to conform with the customary spelling of Hottentot words, when what is required is to conform with the spelling of a geographical name'. Zitzikamma, now, is a place-name in both English and Afrikaans. To what it might correspond in Hottentot is a different question. Such a name, long established in public use, assumes importance in legal documents like title-deeds to land, in which original spellings have to be retained.

To take a different kind of example: In the peculiar circumstances of modern Antarctic exploration a geographical name becomes established as soon as it is given; for by it all specimens collected, samples taken, and observations made are at once localized in the records of expeditions. Collections are subsequently sent for study to specialists in various parts of the world. The results of their work are eventually published in scientific journals or in voluminous technical reports of exploration. Type-specimens and samples are finally deposited in widely scattered museums and institutions. And everywhere the original locality-names are carefully preserved as the only links between the materials brought from Antarctica and the exact places from which they were brought. Subsequent change of these names, to be effective, must be applied systematically along the whole chain of reference, which is impossible.

The custodians of geographical names are therefore under an obligation to science not to change established names for any but urgent reasons. There is too the further obligation of giving wide publicity to necessary changes.

The advance of exploration is one of the few things that make the revision of geographical names necessary from time to time. When it became known that Victoria Land, in the Canadian Arctic, was an island, it was appropriate to change its name to Victoria Island. This change, in turn, affected the name of South Victoria Land, in the Antarctic, for which Victoria Land (its original name as given by Ross) thenceforth sufficed. Such revision is as necessary as the elimination of synonymy from biological nomenclature, but it should be undertaken with extreme caution, and should never be pressed too far towards system.

To take a third kind of example: Arab and Malay sailors, like Portuguese, Dutch, and British sailors, have spread geographical names in their languages far beyond their own countries. In Indonesia we have conventionally used many Malay names of natural features. Being of wide currency some of them are of more use to seamen than the corresponding native names in the local languages. Some of them, however, have been replaced on British charts by Dutch forms, while others, which were not accepted by the Dutch, are disappearing from our charts. The loss of a name like 'Serwatty islands' is as regrettable as the eclipse of the name 'Sandwich islands'. I suggest that no name in English conventional use should ever be replaced by a name in foreign conventional use.

It is difficult to see why we should, as some authorities seem to wish, impose on the gazetteer a kind of regularity that we would never expect to find in the telephone directory. It would doubtless make the clerical work of the telephone authorities much easier were the people who call themselves Philips, Philipps, Phillips, and Phillipps all compelled to spell their names in the same way. But they would certainly object to doing so. Geographical names (as distinct from toponyms), particularly in unhabited country, have little defence but that of our respect. They exist in the service of geography, not in that of clerical convenience. What regularity they have is related to the conventions of the times when they were given; and on their irregularity, indeed on their irrationality, depends much of their effectiveness.

Pronunciation of names. The main purposes of writing are to record speech and thought. It is probable that all alphabetical and syllabic systems of writing were phonetic when first devised and applied. It is well known that the pronunciation of words changes more or less rapidly, but in general much more rapidly than their spelling, unless reform in orthography keeps pace with phonetic change, which it rarely if ever does. English spelling lost its flexibility when Johnson's dictionary established the canon in 1755; but even in Johnson's day there were two camps, the one maintaining 'that there was some general way of pronouncing certain vowel-symbols, or combinations of symbols in English, which, if adopted as the basis of transliteration, might ensure a correct pronunciation of Oriental words by non-Orientalists; the other despairing of extracting any order out of our chaotic vowel-combinations, and founding their system of transliteration on the more regular Italian or French method of vowel-pronunciation' (*sic*, meaning vowel-representation). That is, there was a school which advocated Feegee and another which favoured Fiji. The orthography of the former school was widely used for geographical names up to the earlier part of nineteenth century, and is still often used as a guide to pronunciation.

Towards the middle of the century fundamental research on the nature, recording and transmission of sound, including that of human speech, was in progress. The work then being done by such men as Sir Charles Wheatstone and A. J. Ellis made possible our use of telegraphy and modern shorthand. In 1863 Bell produced his Visible Speech Symbols, and so drew the line sharply between the orthographical conventions of written languages and the exact notation of spoken sound. His symbols were adopted by the Philological Society and the Clarendon Press for use in *A New English Dictionary* (Oxford, 1884–1928), and have remained in use in supplements to, and abridgments of, the dictionary which have since been issued, although they are now superseded in phonetics by the symbols of the International Phonetic Association (1886).

English has for long had its advocates of reformed spelling, and if Bernard Shaw had properly appreciated the vigorous campaign of Alexander John Ellis a hundred years earlier, he might not have bequeathed us his own curious scheme. Phonetic alphabets, whether Roman or consisting of special symbols, are not adapted to the writing of geographical names in English context. There is often, nevertheless, a need to indicate the pronunciation of a difficult name. The only scientific way of doing so is to write it in International Phonetic symbols, which are incomprehensible to the public. Sundry English conventional systems of spelling which give 'imitated pronunciation' have therefore been developed for auxiliary use in gazetteers and grammars. Notable amongst them are those of Chisholm's *A Pronouncing Vocabulary of Geographical Names* (1912), of *The Times Survey Atlas of the World* (1920), *Webster's Geographical Dictionary* (Springfield, Mass., 1949) and *Chambers's World Gazetteer and Geographical Dictionary* (Edinburgh, 1954). The last two bring out the differences between English and American pronunciation.

The imitated pronunciation given in such publications is not necessarily correct, for pronunciation is neither fixed nor uniform within a language, and the correct pronunciation of innumerable names is difficult to ascertain, even from natives. The names Dongara and Pilbara, for instance, are commonly stressed on the second syllable in eastern Australia, but on the first in Western Australia where the places are situated.

The best statement on the pronunciation of geographical names is that of Professor Lloyd James.[1] Great Britain is fortunate in having a pronouncing authority in the British Broadcasting Corporation, which takes considerable trouble over proper names, personal and geographical, British and foreign. It has collected a large number of foreign geographical names and presented them in International Phonetic script, both according to their correct native pronunciation, and to pronunciation suitable for English broadcasts.[2] The Columbia Broadcasting Corporation has done corresponding work [3] for the American public. In both instances the principle followed has been to acknowledge the impracticability of trying to use the true native pronunciation of foreign names, and the necessity of providing a recognizable pronunciation which will not impede the natural flow of the respective varieties of spoken English.

A marked change in opinion has taken place of recent years with regard to the importance of the pronunciation of foreign geographical names. In 1884 d'Abbadie wrote: '.... les noms servent chez nous à identifier les lieux, et l'immense majorité des lecteurs ne se soucie guère de les prononcer. En changeant l'orthographe d'un mot, on lui ôte sa physionomie et l'on empêche de le reconnaître.' Again, in 1897, Ritter von Steeb wrote: 'Darüber sind aber alle einig....dass die Schreibung der geographischen Namen wichtiger sei als die vollkommen genaue Aussprache derselben'. We are not now so sure about it, since so many listeners are accustomed to hearing foreign names correctly pronounced in wireless broadcasts, and something like a revival of interest in the learning of foreign languages is in progress.

[1] *See* Broadcast English, vi, *London*, 1937, pp. 1–16.
[2] Op. cit., footnote 3, p. 27.
[3] *See* World Words, by W. Cabell Greet, *New York*, 1948.

It proves necessary, from time to time, to prepare **lists of geographical names** for contingent purposes, when the question of their alphabetical arrangement at once arises. In spite of the fact that alphabetical order is well settled in European alphabets, the arrangements of words in alphabetical order has its problems, as every indexer knows. Am I to list the ports under Port, or under the proper element of each name? Port Jackson, or Jackson, Port: and if the former, am I to list them from Port A——— to Port Z——— before I go on to Portarlington, or does it come into its alphabetical place amongst the Ports A———? What do I do about 'The Skaw'? Is it to be Skaw, The:? These are not questions to which there is a final answer, for different systems of listing suit different purposes, and of several possible methods, none is necessarily wrong. The Admiralty in its Sailing Directions, for instance, lists names beginning with Saint, Sainte, Sankt, San, Santa, Santo, São, etc., all under a simple S., which is of course explained in an appropriate place. It is left to the reader to expand the word, unless he turns from the index to the text of the volume. The following remarks may, however, be helpful.

If our list consist of names from all over the world, or from a great part of it, English alphabetical order is best, in which foreign letters carrying diacritical marks are treated as if they had no such marks. An *ñ*, for instance, although it is so printed in our list, is treated simply as *n*. But if the list be of names from one country, and that country use a Roman alphabet, we do best to list the names in the precise order of the alphabet under consideration. This may produce unexpected and un-English results, but of the very kind which the user will have to accept from foreign sources and in the foreign dictionary, and it will make our list directly comparable with corresponding foreign lists.

Every name should be listed under its proper element, but if several things of the same name appear, it may be possible to distinguish between them, as with Fall River (the city) and Fall river (the river). In a list in English order no useful purpose is served by listing definite forms of names under *the*; but the *le* or *la* or other article of many Romance names may not only be proper to them, but may even be written together with the name as one word. The *el* and *al* in Arabic names may be troublesome, for Arabic *el* (the article) may conflict with Somali *el* (which means a well); or Arabic *al* (the article) with Arabic *āl* (which means family).

Compound names of natural features, in which the specific element follows the generic, may profitably be inverted for listing. Port Jackson, for instance, is the name of a harbour and of nothing else. There is no place called Port Jackson. It can therefore appear in a list as 'Jackson, Port:'. But Port Lincoln is the name both of a port and of a settlement on it. The proper name of the settlement is Port Lincoln in full, and might equally well be written Port-Lincoln, or Portlincoln. Such a compound proper name should not be inverted. But, if particulars of the port be given under the name of the settlement, a cross-reference from 'Lincoln, Port:' to 'Port Lincoln' is necessary. Without the inversion of the compound names of natural features, a big list or full length gazetteer would comprise a number of secondary lists, also in their own alphabetical order, under each geographical term of importance, like the secondary lists which appear under Mac and St., in the telephone directory.

The inversion of certain compound names can be avoided only at the expense of the margin of the printed list. With the proper element brought to the front, the column of type can be as near the left-hand side of the page as desired; but if the following alternative scheme [1] be used:

London
City of London
County of London
Londonderry

a correspondingly wide margin is required on the left-hand side of the page.

It may also be asked how to treat names like East London (South Africa), Old Providence, North Shields, South Shields, etc. As East London has nothing whatever to do, geographically, with London, nor Old Providence with Providence, and as North and South Shields can hardly be regarded as parts of one place, these names are all proper as they stand. But the names of sets of neighbouring villages beginning with words meaning Upper, Middle, Lower, or Inner, Outer, or Great, Little, etc., often refer to places so close together that they are really parts of the same place. It is advantageous to invert them, and so bring them together in the list.

[1] cf. *Geogr. J.*, 91 (1938), pp. 559–560.

CHAPTER VII

PROCEDURE RECOMMENDED

The P.C.G.N. *Principles of Geographical Nomenclature* (October, 1954) were drawn up for the guidance of official organizations which produce charts and maps of large scale. To the everyday problems of biologists, historians, teachers, journalists, editors, publishers, broadcasters, and others, they do not necessarily offer the most satisfactory solution. But before considering wider principles which might meet the more general need, it is necessary to state what is possible, and to demonstrate what, in my opinion, is required.

There are very few things which can be done with the name of a foreign geographical feature for which we need an English written form. We may take it directly, if it be written in the Roman alphabet; we may render it letter-for-letter from a foreign Roman alphabet or from any non-Roman alphabet; we may render it sound-for-sound, either by ear from any language, or by transcription from any alphabet or non-alphabetical script; we may translate it, if it have meaning; or, if it be a compound name containing a geographical term and a proper noun or adjective, we may turn it into an English name by partial translation.

The last is difficult, and satisfactorily done only by persons with a considerable knowledge both of English and the other languages involved, and of geography. The Greek name Κορινθιάκος Κόλπος (Korinthiakos Kolpos) for instance, is literally 'Corinthic Gulf', but might equally well be rendered 'Gulf of Corinth' or 'Gulf of Korinthos'; to which most would say, 'Let us have Gulf of Corinth if it is to be English and Korinthiakos Kolpos if it is to be Greek'.

If a foreign place-name (toponym) under consideration be that of a centre of population we do wrong to tamper with it at all, except for necessary transliteration or transcription. In rendering foreign names it is most important not to impede ready identification of the thing named by producing unfamiliar variants of the name, and just as important not to break the continuity of our records by upsetting alphabetical order if it can be avoided. The example of Korinthos above would transfer a well-known name to a new place in English alphabetical lists.

It is also important to prevent the growth of linguistically hybrid names like 'Ak river' (from Ak Su, which means 'white river'), and of pleonasms like Lake Windermere (lake and mere expressing the same meaning). Pleonasm is particularly common on some Russian maps, but names like 'Pentire Head' are not unknown on English maps, 'pen' meaning head.

It is also well to remember that important geographical names have continually to be printed, copied, telegraphed and written on maps and charts, and the more simply they are rendered (without sacrifice of essential distinctions), the better for all who have to use them.

The intelligent person at home, reading the newspaper or consulting the atlas, wants geographical names which he can recognize, or can at least identify with little trouble; and he wants them written so that he can say them. Such names as Płock and La Jolla (pronounced Pwatsk and La HO-ya respectively) are phonetically misleading to an Englishman, who thinks, and not unreasonably, that foreign words should be pronounced as they are written. Whether he pronounce them correctly or not is relatively unimportant when speaking English. An announcer broadcasting the news will prefer to use a form of name which will immediately be recognized as something recently seen in print. Topical books, small atlases, and maps on small scales will succeed best if they present the easy and familiar. Established English and conventionally English geographical names, and broad phonetic renderings of less familiar names seem therefore to be the most suitable for popular use.

One has only to cross to Belgium or France to learn that English names are not understood abroad. British maps and charts for the use of our soldiers, sailors, and travellers abroad, must present what will be of use there. Such maps are commonly on large scales, and the names on them should if possible be so written that they will be understood when read aloud, or the way in which they are written must be so well explained that their intelligible pronunciation can be mastered after a little practice. The British scholar or linguist abroad will probably use the best foreign maps he can obtain and will give little thought to the possible ways of rendering the names on them; but the same scholar, publishing the results of his researches, will strive to render the names in the English alphabet in a way that will enable his fellow scholars to discern the originals through the transcriptions. And British librarians, concerned with the literature of all countries and all times, and with the necessity of transcribing all scripts into our own before names and titles can be arranged under the English alphabet, will reject phonetic rendering if it be misleading for their purpose. What they desire is reversible transcription, even if it be of a mechanical kind, provided that it maintain continuity in cumulative catalogues. British scholarship demands that, for scholarly purposes, foreign names be transcribed by rigorous methods.

From the beginning, therefore, we encounter the conflicting claims of the spoken and the written word, the ordinary man and the scholar, the foreign language and our own. Now, we say of books that 'This one is in French, that in English', and we do not hesitate to take a dictionary when we use the French book. We should be prepared to do the same with maps, for they are documents which concentrate information more compactly than books. We can read a large-scale map in a foreign language only with the aid of a glossary. Except for the necessary transcription or transliteration of the names on them, our large-scale maps, meant for use abroad or for technical use, are in the languages of the countries they represent. Such of our small-scale maps as are for popular use should, I consider, be in English; that is, we should make the fullest use on them of our stock of established English and conventionally English names.

In the peculiar circumstances of our times, the readiest and most abundant source of geographical information about a country is likely to be in its official maps and charts; but they are, even at the best, a provisional source which should be regarded as subject to check. In a rapidly changing world we may need our information quickly, and, as regards geographical names, we now often take it, but not, I hope, uncritically, from the latest official map of the country under consideration.

This constrains us to the treatment of foreign names on **a basis of political, not of linguistic unity.** For example, written Arabic is the same language throughout the Muhammadan world, though the spoken Arabic of one part may be almost incomprehensible in another. The literary unity of Arabic corresponds to the wonderful literary unity of Chinese. There is every reason, linguistically, for transliterating Arabic according to one system throughout. For geographical purposes, however, it is much more expedient to accept the different results of Spanish, French, Italian, and English systems for names in Spanish Morocco, French North Africa, Libya, and the Arabian Peninsula respectively, if we want to find places on the maps of those countries, or if we want our mail to be delivered in them, though the question of mail is not elsewhere so simple.

Another important contemporary source of information about changes abroad lies in the Summaries of World Broadcasts issued by the British Broadcasting Corporation, which are obtainable on payment. The geographical names which appear in them are not necessarily in foreign official form even if officially written in the Roman alphabet; and those from countries which use non-Roman alphabets or scripts may appear in mere phonetic rendering.

There remains the question of an English alphabet into which to transcribe and transliterate foreign names, for our common English alphabet will not do. Our alphabet is imperfect and our use of it both conservative and inconsistent. According to Monier-Williams we use as many as fourteen ways of representing the sound of *e* in *be*. To achieve regularity in the rendering of names from other alphabets and scripts we are therefore constrained to use the English letters conventionally. The long history of the development of a suitable conventional alphabet has been outlined in the section on English Usage above (pp. 72–76) and it should be evident that the alphabet recommended by the P.C.G.N. represents the sum of English experience in the handling of foreign geographical names.

Within the limitation of treatment according to political rather than linguistic unity, the following cases have to be considered: those of the names of places and geographical features situated

(*a*) beyond territorial limits;

(*b*) in territory of disputed ownership;

(*c*) within the territory of federated states, (i) in the federal official languages, e.g. Great Russian, (ii) in the languages of each of the several federated states, e.g. Ukrainian, Estonian, and (iii) in languages subordinate to the last, e.g. Ossetian;

(*d*) within the territory of independent states, (i) in the official language and script, and (ii) in languages elsewhere normally written in a different script, e.g. Arabic in Israel, Bulgarian and Turkish in Greece;

(*e*) within Colonial, Mandated or Trust territories, (i) in the official spelling or script, and (ii) in vernacular spelling or script, e.g. Fiji.

It must be freely acknowledged that (*b*), (*c*) ii and iii, and (*d*) ii present problems which have not yet been satisfactorily solved. The reader is invited to consider the question of Berber names in North Africa, or of Fula names in West Africa.

Early in 1942 I suggested a dual scheme for the writing of geographical names, of which the particulars were exposed for criticism in the *Geographical Journal*.[1] Subsequent discussion has not convinced me the need for any radical modification of the proposals in principle, and I now submit them again, after slight revision, for public consideration.

Dual system for the writing of geographical names

1. By geographical name shall be understood a proper name or geographical expression which serves to distinguish a particular geographical entity from all other things.
2. Explorers, travellers, and surveyors who propose names for geographical features not previously known by name to any people should not suggest commonplace or frivolous names, and should publish the names proposed only with the approval of the government in recognized control of the country or region concerned.

[1] 100 (1942), pp. 177–181 and 245–256.

3. Explorers, travellers or surveyors, when recording unwritten geographical names in the field, should endeavour to do so by the scientific methods now employed for the phonetic recording of speech, failing which, a record in the symbols of the International Phonetic Association should be made, or in the vernacular alphabet if there be a suitable one in use.
4. Those concerned with the transmission to Home authorities of information about established geographical names abroad, should endeavour to render local pronunciation in the symbols of the International Phonetic Association, and to supply both native and official written forms, should they differ, as written by native scholars, in the native and official scripts.
5. In technical prose-context, in reference atlases, and on plans, maps, and charts on the scale of 1:1,000,000 and larger:

    (*a*) The name of any geographical feature which lies beyond the territorial limits of recognized government or of any geographical feature or region of international extent, or of any country, which has an English or conventionally English name, shall be that name.

    (*b*) In countries which officially use varieties of the Roman alphabet all geographical names but those of the kinds mentioned in (*a*) shall be adopted without change in their official forms.

    (*c*) The non-Roman letters in geographical names in countries which use partly-Roman alphabets may be transliterated into those of the conventional alphabet of the P.C.G.N.

    (*d*) Geographical names, except of the kinds mentioned in (*a*) above, in countries which officially use non-

Roman alphabets, non-alphabetical script, or auxiliary systems of romanization, shall be strictly rendered in romanized form according to systems of transliteration or transcription recommended by the P.C.G.N.

(*e*) In historical context the name of any part of the world, geographical feature or place shall be that which it bore in the region and language concerned, at the time under consideration; or, a received transliteration or English transcription of that name, were it originally written in a non-Roman alphabet or script.

(*f*) In archæological context the name of any site shall be that by which it is conventionally known to English scholars.

(*g*) To ensure its identification, the historical name of any important place or geographical feature, or of any archæological site, may be followed in brackets by the corresponding English, conventionally English, international-postal or modern official name (strictly transcribed or transliterated from any non-Roman alphabet or script).

6. In speech, in popular prose-context, in elementary atlases, and on maps, charts, and diagrams on scales smaller than 1:1,000,000.

(*a*) The name of any part of the world, country, geographical feature or place which has an established or conventional name in the English language shall be that name.

(*b*) The name of any other geographical feature or place shall be that used by the government in recognized control of the country or territory in which it is situated, but without diacritical marks, if the name be officially written in a form of Roman alphabet, and a strict transliteration or transcription, according to a system recommended by the P.C.G.N., if it be officially written in a non-Roman alphabet or non-alphabetical script;

provided that, should a foreign form in an official Roman alphabet or in English transliteration or transcription be phonetically misleading to the English public, its pronunciation be given in brackets, in geographical texts and indexes, according to the principles used by the B.B.C. in its ' Broadcast English' series. (The alphabet used by the P.C.G.N. being conventional, is only crudely phonetic, and its possibilities for imitated pronunciation are therefore limited. Sometimes it is adequate, e.g. Niš, pronounced Nish; Cetinje, pronounced Tset-in-ye).

Whether we like it or not, geographical names have to be considered from two opposed positions, that of our own language, and that of the rest of the world.

The clear recognition of the forms of geographical names which are most acceptable as good spoken English, and of the written forms which are best suited to our closer dealings with other countries, will help to maintain the clarity, precision, and character of our language, and the dignity of our relations with peoples abroad who do not speak it.

CHAPTER VIII

INTERNATIONAL USAGE

The extent of international understanding—it can hardly be called agreement—on the use of geographical names is very limited. The *General Instructions for Hydrographic Surveyors* of the Admiralty, of 1862, said that 'those [names] which have been stamped upon places by the first discoverers are held sacred by the common consent of all nations'.

Some of the names given by English discoverers have been treated with great respect by Norway in Svalbard[1] and by the Argentine and Chile in southern South America, where Beagle Channel, for example, is now Canál Beagle; but Cook's name 'Sandwich Islands', in another region, has not been allowed to survive. Australia, on the other hand, has been at pains to restore the original French names to that part of the coast of South Australia discovered and charted by the French.[2]

An Australian inter-state conference held in Melbourne in 1884 acknowledged the priority of native names. The seventh International Geographical Congress (Berlin, 1899) decided that native names, established as accurately as possible, should prevail over others, a principle supported by the Royal Geographical Society (1901), the Colonial Division of the German Foreign Office (1903) and the United States Geographic Board (as it was then called). The same Congress also decided that, where there are no native names, or where they cannot be established with

[1] The Place-names of Svalbard. *Skrifter om Svalbard og Ishavet*, 80 (1942).

[2] See *Proc. R. Geogr. Soc. Australasia, South Austr. Br.*, 14 (1912–1913), p. 20.

certainty, the names given by first discoverers should prevail. Hence, perhaps, our loss of the Sandwich Islands.

International discussion of appropriate ways of writing geographical names, has been going on more or less actively for the past two hundred years, though with little real advance. It has followed, in the main, three lines of recommendation, one for the use of a universal script, one for that of a mixed alphabet, and another for that of a Roman alphabet.

In 1820 the Comte de Volney offered, through the Institut de France, a prize of 1,000 francs to the inventor of the best **universal alphabet** for use in transcription ('alphabet transcriptif Universel'). Interest in the possibility of universal script is as old as that in the possibility of universal language, and numerous attempts have been made to produce them, but the Institut de France showed admirable restraint towards the award of the Volney prize. In 1848 Alexander John Ellis brought forward proposals for a universal alphabet which he called the Ethnical Alphabet, or the Alphabet of Nations, for which handsome type was designed and cast.[1] Ellis's scheme was ingenious, the strange letters he proposed were attractive in appearance, but this alphabet aroused no more response than his sustained proposals for the reform of English spelling. At the sixth International Geographical Congress (London, 1895), Ricchieri had maintained that geographical names should be written in an alphabet of their own, which could be developed by an international commission. This encouraged Garnier to produce his elaborate universal alphabet (1899), but it too has found no response. Garnier, nevertheless, was posthumously awarded the Volney prize,[2]

[1] Ellis, A. J. The Ethnical Alphabet, *Bath*, 1848.
[2] Garnier, C. T.R.G. Méthode de Transcription Rationelle Générale des noms géographiques. *Paris*, 1899.

and Wharton considers his work to have produced 'a system for geographical purposes which has great value'.

Mixed alphabets, with one exception, have found as little favour as universal alphabets. In 1849 Sir John Herschel recommended to the Admiralty, as an alternative to Ellis's Ethnical Alphabet, an alphabet consisting of Roman and a few Greek letters, for the writing of native names. It was quickly rejected in practice in favour of Washington's alphabet. In 1855 Lepsius produced his first full account of his General Linguistic Alphabet,[1] which, because it consists of both Roman and Italic letters, must be regarded as a mixed alphabet. His work had far-reaching effects. It was accepted at once by the Church Missionary Society, and, in modified form, was eventually accepted also by the Clarendon Press for use in the translations of the 'Sacred Books of the East' (1879–1904) '. . . in these valuable books intended for the use of English-speaking Orientalists (including natives of India) throughout the civilized world, we have our common sounds of 'ch' and 'j' (equally common in Indian languages) represented by *k* and *g*. The consequence is that such names as Chandra-gupta, Sach-chid-ānanda, are spelt *K*andra-gupta, Sa*k*-*k*id-ānanda, while Jaina is *G*aina, and—the italics often being omitted by printers in quotations from these works—a false pronunciation of important proper names is being everywhere propagated' (Monier-Williams). This shows clearly one of the disadvantages of mixing different styles of Roman letters.

In 1913 Eserskey recommended a mixed alphabet consisting of Roman and Cyrillic letters for the writing of geographical names,[2] which has not found favour, but

[1] Lepsius, C. R. Das Allgemeine linguistische Alphabet. *Berlin,* 1855.

[2] See *Atti X Congr. Internat. Geogr., Roma,* 1913 (1915), pp. 1515–1524.

which may be worthy of re-examination in relation to the possible development of an internationally acceptable form of Roman alphabet. In 1927 the International Institute of African Languages and Cultures (now International African Institute) introduced an alphabet consisting of Roman letters and of International Phonetic symbols,[1] for the writing of African languages, which, after severe trial, has proved to be well adapted to its purpose. It is now in established use in many parts of Africa.

Romanization, in the strict sense, is the reduction to writing, in Roman letters, of languages formerly written in other alphabets or scripts. Turkish, formerly written in the Arabic alphabet, was romanized in 1928. Because the Arabic alphabet does not fit the sounds of Persian it is often said that Persian ought to be romanized; and there is occasionaly reference to the desirability of romanizing Urdu. A language may be said to be romanized when its own speakers habitually write it in a Roman alphabet in preference to any other alphabet or script formerly used.

Several Oriental countries which officially use other scripts have also developed official systems of transcription of their own script into Roman letters for auxiliary use, particularly in their dealings with Western countries. Egypt, for instance, publishes maps in Arabic and also in Roman transliteration of Arabic. Japan regularly publishes maps in Japanese, but has experimented with different systems of romanization and has issued maps in several Roman transcriptions of Japanese. The examples of Japan and Egypt are of transliteration and transcription rather than of romanization. International romanization is an expression often heard, though generally misused. Taken strictly, it would mean the writing of all languages in an internationally accepted form of Roman alphabet.

[1] *Memorandum* 1. Practical orthography of African languages. Revised ed., *London*, 1930.

Geographical names written in or transcribed into forms of the Roman alphabet are comprehensible to the eyes, if not to the ears, of educated readers throughout the Western world, no matter what or how many diacritical marks may be used in a particular Roman alphabet. That is to say, names written in Roman letters are internationally acceptable throughout the Western world. Though this is far from meaning 'throughout the world', it gives the Roman-alphabet forms a long advantage.

The history of Western efforts to secure the general acceptance of the Roman-alphabet forms is long, and has two important aspects : that of the efforts of western nations to regulate the spelling of geographical names in their own Roman alphabets; and that of the efforts of internationalists to find internationally acceptable spellings of geographical names in Roman letters.

The history of English efforts to write foreign geographical names in a rational way has been outlined above in the section on English Usage. It is most likely that every civilized nation has a history of corresponding effort. In spite of Astley's work, we ourselves were by no means the first to try to solve the problem; nor did it necessarily present itself simply as a problem of how to write geographical names.

To the Spaniards in the Americas and to the Portuguese in Africa and south-eastern Asia during the sixteenth century, it became a problem of how to reduce native languages to writing in Roman letters. Its solution was undertaken by the Sacra Congregatio de Propaganda Fide (Rome) which produced, according to Wharton, 'a magnificent series of . . . primers which are an ever memorable monument of the broad spirit in which the Sacra Congregatio de Propaganda Fide has viewed its duty. It has moreover in its *Tipografia Polyglotta* and in its college continued to keep before the public two ideals, the value of knowing the language of those you convert and the necessity for your own study and the instruction of others of a scheme of transliteration of some sort'.

The first International Geographical Congress, held at Antwerp in 1871, decided to recognize the national official forms of geographical names in all countries which used the Roman alphabet, and the Universal Postal Union, formed in 1874, has consistently upheld that decision ever since. The Union, engaged in identifying and registering all the post-offices in the world, has found it expedient to accept romanized lists of post-offices as supplied by the adherent countries, with results better than might have been expected. The simpler European alphabets, and the simpler conventional systems of Roman transcription found most favour, and the *Dictionnaire des Bureaux de Poste* exhibits no more than a reasonable proportion of diacritical marks and other orthographical devices. Every place with a post-office now has a form of name for international use, in the Roman alphabet.

The action of the Antwerp Congress of 1871 evidently stimulated interest in national **conventional Roman alphabets** for the writing of names of places in colonial territories, for, in 1876, the Sociedad Geográfica de Madrid made the first modern attempt to produce such an alphabet. The Third International Geographical Congress (Venice, 1881) invited geographical organizations to study the question of international orthography and to come to agreement about it. The response was the development of four more national conventional Roman alphabets: our own R.G.S. System in 1885, a French alphabet in 1886, a German in 1888, and a Belgian in 1892.

The Ninth International Geographical Congress (Geneva, 1908) appointed a commission to study the writing of geographical names, but in 1909 an independent committee, appointed to organize the preparation of an international map of the world, held its first meeting, in London. The International Map Committee resolved to accept officially romanized forms of names, and invited countries which used non-Roman scripts to develop official systems of romanization. At the tenth International Geographical Congress (Rome, 1913), the commission appointed at Geneva did no more than recommend that the question be referred to the International Map Committee, which was then holding its second conference, in Paris; but the Geographical Congress brought discussion to an end by accepting the resolutions passed by the International Map Committee in 1909.

The **Roman alphabet rule,** that is, the principle of accepting the official forms of foreign geographical names if they are officially written or officially transcribed into forms of the Roman alphabet, thus became established. It was accepted by the International Hydrographic Conferences of 1919, 1926, and 1929, and was confirmed by the fifteenth International Geographical Congress (Amsterdam, 1938) which recommended for international use the forms of geographical names appearing in the *Dictionnaire des Bureaux de Poste.* The rule marks the greatest measure of international accord so far attained on the question. But it is particularly to be noted that it does not establish truly international usage. It has established no more than a measure of international acceptance of national usage, for the names in the *Dictionnaire* appear in the forms supplied by national postal authorities.

Attempts have been made to devise a Roman alphabet suitable for use throughout the world, and several such alphabets have been offered for the writing of geographical names. One of the first was that of Sir Robert Schomburgk, who applied Sir William Jones's principle to the writing of the languages of British Guiana, and made proposals for 'a uniform system to reduce unwritten languages to alphabetical writing in Roman characters' in 1848. In 1853 the Chevalier de Bunsen convened a series of alphabetical conferences in London, which led Max Müller to produce a Missionary alphabet in 1854. He based it on the confirmation, by the physiologists of his time, of the analysis of the Devanagari phonetic system by the ancient Sanskrit grammarians. At the fifth International Geographical Congress (Berne, 1891), it was generally agreed that a special alphabet should be developed for rendering names in internationally acceptable form. Köppen thereupon (1893) produced his Roman alphabet modified for international use, but it is clumsy, requires too many diacritical marks for the convenience of any but Croat, Czech or Polish printers, and has been disregarded. For the same reasons an elaborate and highly scientific Roman alphabet devised by Father P. W. Schmidt in 1907 has been neglected by geographers, though it has found response amongst linguists.

By 1913 disbelief in the possibility of international romanization was widespread among geographers, but was overtly expressed in England only by the late Mr. A. R. Hinks (representing the Royal Geographical Society on the International Map Committee) and the late Lord Edward Gleichen (Chairman, P.C.G.N.). Without an internationally accepted Roman alphabet international romanization cannot be achieved, but the different values assigned to c, g, h, j, v, w, x, y, z; the variety and inadequacy of devices for indicating aspiration, palatalization, nasalization and yodization; and the poverty of means for indicating stress, quantity and tone, in western European alphabets, have defeated all efforts to establish one.

This, fortunately, is not the end of the story. For two attempts at **regional romanization** have succeeded. The history of the application, by Europeans, of Roman letters to the writing of the languages of India is a long one, and those interested will find references to the literature of the subject in papers by Max Müller, Lepsius, and Monier-Williams. No firm basis for a system was found until Sir William Jones announced his principle in 1788; and no generally acceptable alphabet was proposed until Sir W. W. Hunter submitted his own to the Government of India in 1870. His system was accepted, though in modified form, in 1875.

Hunter's system of spelling, subsequently called the Hunterian System, differs but little from the R.G.S. system, but permits the use of diacritical marks to indicate long vowels and retroflex consonants (those pronounced with the tip of the tongue curled up and back to touch the palate), which are characteristic of the Indian languages. It is commonly supposed that Indian names have been systematically transcribed into the Hunterian alphabet from the native scripts, which is no doubt true of many; but many others have been written down by ear in the Hunterian alphabet without any reference to their forms in native script. The Hunterian system is a remarkable instance of the successful romanization of names belonging to a great variety of languages. Authorities in Europe seem not to have taken it into account in their efforts to achieve similar results.

The second attempt was made in a very different region. During the nineties of last century the Austro-Hungarian General Staff discovered the inconvenience of using several systems of romanized orthography on the map of a single region. 'Der Kartenbenützer müsste immer erst bestimmen, welcher Nation der betreffende Name angehört und danach die Aussprache der lateinischen Lettern modificieren. Das ist für eine Soldatenkarte zu compliciert' (Ritter von Steeb). It then made a bold experiment in regional romanization, and rendered all names in the Balkan peninsula in an amplified Croat alphabet, which is still successfully, though unofficially, used for rendering Cyrillic names in romanized form.

Now, two things are here to be noted: In the sustained discussions held by geographers between 1871 and 1909, the main objection to the forms of Roman alphabets proposed for international use was to diacritical marks—of which English and American cartographers are still afraid; and I have said above that a quickening of English aural perception seems to have taken place during recent years. I have therefore witnessed, with the greatest interest, the amplification of the R.G.S. II alphabet by the P.C.G.N., in the direction of the amplified Croat alphabet through the acceptance of diacritical marks, and in the international direction through a slight retreat from Sir William Jones's principle with regard to consonants.

The possibilities. Learned Europe in the Renaissance observed the orderly, international custom of writing in Latin, and reduced the names of the more important places and geographical features of its world to Latin form, thus achieving what is again desired: systematic international usage. Unfortunately, Latin has fallen into disuse as an international vehicle except in the Roman Catholic Church and in a few special interests like systematic Botany.

The British Association for the Advancement of Science has carefully considered the suitability of several linguistic media available for accessory use in science, but the notations of music, mathematics, chemistry, and telegraphy remain the only truly international intellectual vehicles.

Just as there is no international Roman alphabet, there is no international language. The proposals that the spellings of names in one language, such as French, be accepted for accessory international use, was rejected at the fifth International Geographical Congress (Berne, 1891). And even before that, the United States Hydrographic Office had faced the whole question of foreign geographical nomenclature in an international spirit (1889), by consulting all the maritime nations in the hope of attaining a uniform system of spelling. This effort failed.

The Roman alphabet rule is only a provisional solution of a fully international problem. The names listed in the *Dictionnaire des Bureaux de Poste* represent a number of national alphabets and present a proportion of diacritical marks, which, though reasonably low, is enough to make the names difficult for many countries to copy in type.

In spite of its great practical convenience to postal authorities throughout the world, the rule falls far short of meeting the needs of geographers, even in countries which officially use the Roman alphabet. For the names listed in the *Dictionnaire* are only those of inhabited places where there are post-offices, besides which geography has to deal with places where there are no post-offices, and with the untold host of names by which parts of the world and the natural features of its surface are known. Enormous numbers of these names have to be transliterated or transcribed into our various alphabets and scripts according to national or academic systems, some of which are admirable, but which, taken together, exceed our needs.

Before any really extensive and authoritative work on geographical names can be done by any international authority, national governments will have to lay its foundations by establishing their own toponyms on the soundest bases of toponymic study and orthography. They might be urged to do so by an international authority, through the production of **national gazetteers** prepared according to agreed international standards. Each national gazetteer would then become a contribution towards a gazetteer of the world, and could be kept up to date by periodical revision. Such a scheme was proposed to the P.C.G.N. by Commander Alun Jones, R.N., in 1945, but was then thought to be far beyond the resources of the time.

It must never be forgotten that nations produce maps and charts primarily for the sake of national, not international convenience. Nevertheless, were each nation to reduce its own geographical names to the best possible order, and then to produce official gazetteers for the territories in its possession, the sources of information would at least become clear. Were each nation to give also the most widely accepted pronunciation of each name in International Phonetic Script, a continuous level of international understanding would at last have been established. That this is not counsel of perfection should be evident from what the British Broadcasting Corporation has achieved through its *Broadcast English* series.

Before any programme of work is actually put into effect it will be necessary to ascertain what organizations are interested in (*a*) **international mapping and charting,** and (*b*) **orthography, transliteration, transcription, and geographical terminology.** In my opinion (*a*) should include the following *geographical bodies*: the International Geodetic and Geophysical Union (IGGU), the International Geographical Union (IGU), the International Hydrographic Bureau (IHB), the World Meteorological Organization, the Conseil Permanent pour l'Exploration de la Mer, and the following *communicational organizations*: the Universal Postal Union (UPU), the International Telecommunications Union, and the International Civil Aviation Organization (ICAO); while (*b*) should include the following *linguistic bodies*: the International Phonetic Association, the International Auxiliary Language Association, the International Committee of Onomastic Sciences (ICOS), the International Federation of Orientalists, the International African Institute, and the following *documentational organizations*: the International Federation for Documentation, the International Library Commission, the International Standards Organization, and UNESCO.

The United Nations, whose Economic and Social Council (ECOSOC) is already interested in cartography, might well undertake the coördination of efforts towards the writing of geographical names in internationally acceptable forms, but should have the close coöperation of the International Committee of Onomastic Sciences for the resolution of linguistic and orthographical difficulties, and of the International Geographical Union for that of cartographic difficulties.

Merely to suggest that geographical names might be rendered in a way acceptable to the whole world is to imply belief that some simple and uniform way of recording them is possible. For the names of extra-territorial waters, of international marine, submarine, and terrestrial features, of geographical limits, and of independent countries, the forms current in a particular language may prove suitable; but those of all places and geographical features within or belonging to independent countries would have to be accepted in the forms preferred by the respective independent governments, or in approved transcriptions of them.

The Roman alphabet is the official script of a very large number of countries, and is in auxiliary postal use practically throughout the world. It is suited to the writing or transliteration of most Indo-European languages and of some of the Semitic languages, but is considered to be unsuited for the writing of certain African languages. The International African alphabet is deservedly finding favour in many parts of Africa, but it is a mixed alphabet, only partly Roman. In the experience of the P.C.G.N., names in Bantu, Negro, and Nilotic languages can be transliterated from the International African alphabet, with some loss of phonemic distinction which may impede the immediate identification of some places; the transliterated forms are considered, nevertheless, to be suitable for maps though not for gazetteers.

The development of a Roman alphabet with relatively few diacritical marks, into which non-Roman alphabets might easily be transliterated and non-alphabetical scripts be transcribed, does, indeed, seem to be a practical possibility if nations be willing to pay the 'international rate of exchange' on transliteration and transcription. This might mean the acceptance of familiar symbols used in unfamiliar ways; and of loss of precision in some instances but of gain in others. For instance, we would not be asked to accept Jork for York; but we might have to accept Jenisej for our more familiar Yenisei. It is not for me to recommend an alphabet, but I have no hesitation in recommending the Hunterian, the amplified Croat, and the P.C.G.N. alphabets as the sources of appropriate symbols for an international Roman alphabet. They should be considered with due regard for such historical proposals as that of Lepsius, that of Max Müller, and that of Garnier; and for the more recent proposals of Köppen, Schmidt, Esersky, and others.

Meanwhile, I suggest the following practical measures which might go far towards meeting the immediate needs of our times:

(1) That we endeavour to decide what geographical entities are so widely known as to have names in the leading languages of the world.

(2) That we continue to recognize the names of all places and geographical features within the territories of independent states in their official forms, and accept for international use those written in national Roman alphabets.

(3) That we recommend the transcription and transliteration of the official forms of names written in all other national scripts and alphabets into a conventional but internationally acceptable Roman alphabet.

As regards (1), I submit that it will be appropriate to recognize only the present names of extra-territorial seas, of major international geographical features and regions, and of independent countries. A map carrying these names, and indicating those countries in which the Roman alphabet is the national official script might then be prepared in one language, say French. As a basis for such a map there are three so-called 'international' maps already in existence: the *Carte international du Monde au Millionième* (incomplete; sheets issued by national governments); the *World Aeronautical Chart* (incomplete, issued by ICAO with the coöperation of national governments); and the *Carte générale bathymétrique des Oceans* (issued by the IHB). As regards (3), such an alphabet might well assist in that lowering of barriers between related languages which Rundle recommends as a practical measure within our power.[1]

Should it prove that an amplified Roman alphabet is not acceptable throughout the world for such auxiliary use as the writing of geographical names, the question of a script appropriate for truly international use should be seriously examined. The Arabic and Cyrillic alphabets, and the Chinese and Devanagari characters, though not used over so large an extent of the world as the Roman alphabet, are nevertheless used by enormous numbers of people. Persons accustomed to reading foreign maps may agree that the Arabic alphabet and the Chinese character have much to recommend them for cartographic use, as they crowd the map less than do the Roman and Cyrillac alphabets.

It may prove that the problem is not yet capable of more than regional solution; that, side by side with the use of an amplified Roman alphabet over large parts of the world

[1] Rundle, S. Language as a Social and Political Factor in Europe. *London,* 1946.

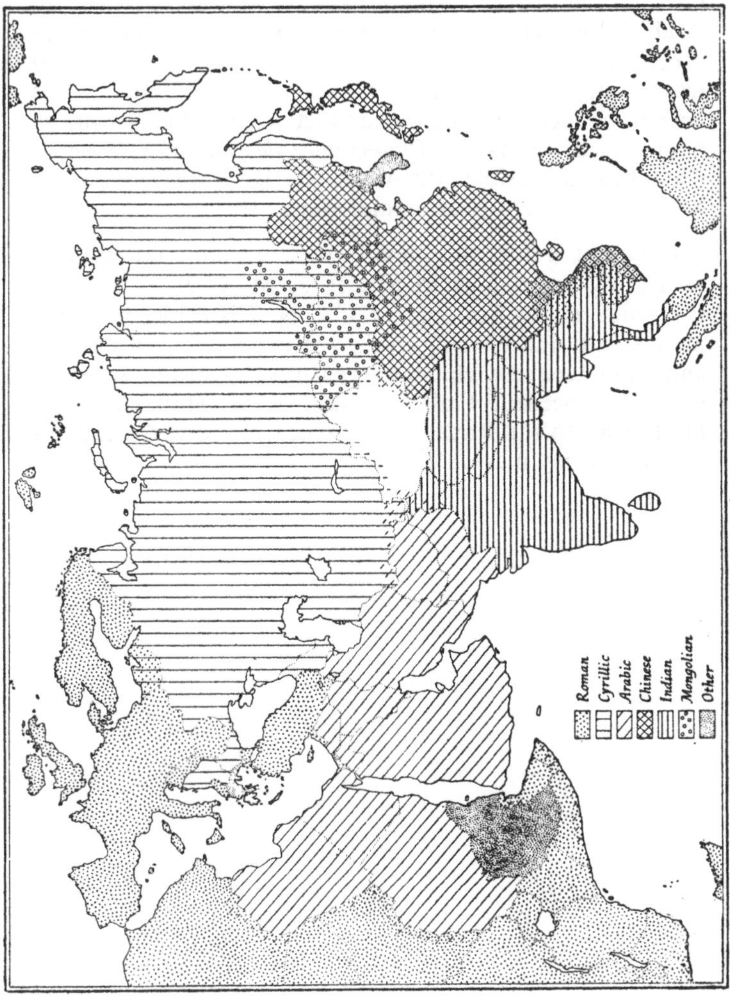

FIG. 5.—*Systems of writing in use in the Old World. Throughout the New World (the Americas, Australia and Oceania) the Roman alphabet is in official use.*

must go the use of a generalized Cyrillic and a generalized Arabic alphabet, and of generalized Chinese characters, in other parts of the world. The need for transliteration and transcription between them would still remain, and would have to be met, were the world to have a list of geographical names as useful as the *Dictionnaire des Bureaux de Poste.* That this is feasible, the work of the P.C.G.N. on the writing of African place-names stands as proof.

I cannot bring this small work to an end without expressing my thanks to Mr. P. J. M. Geelan, who has succeeded me as Secretary, P.C.G.N. His criticism of the manuscript, and his help in the correct copying of Chinese and Sino-Japanese characters ensured precision where it was most needed. I also thank Mr. G. Holland, Chief Draftsman to the Royal Geographical Society, for the fair drawing of my maps.

AUTHORITIES

(Other than those mentioned in the text and footnotes)

ADMIRALTY. A Manual of Scientific Enquiry. Ed., Sir J. F. W. Herschel. *London*, 1849, pp. 441–444.

—— Eskimaux and English Vocabulary. Ed., Capt. J. Washington. *London*, 1850, pp. ix–x.

—— General Instructions for the Hydrographic Surveyors of the Admiralty. *London*, 1862, p. 25.

—— Circular to Admiralty Surveyors. 'Orthography on Admiralty charts'. 10th August, 1884.

—— A system of orthography. Adopted by the Admiralty for sailing directions and charts. [*London*, after 1884 but before 1905.]

—— Instructions for the spelling of place-names in foreign countries. Naval Staff, Intelligence Division. [*London*], 1917. Reissued by the Clarendon Press, *Oxford*, 1918.

—— Index Nauticus. British Isles. *London*, 1920.

—— Professional Paper No. 11, Hydrographic Dept. 'Glossary of Terms used on Admiralty charts and in associated publications.' *London*, 1953.

AÏTOFF, D. De la transcription des noms géographiques. *Ann. Géogr.*, 6 (1897), pp. 14–22.

AMERICAN LIBRARY ASSOCIATION: Committee on transliteration of Slavic languages. 'Catalog rules'. *Boston, Mass.*, 1908.

ANON. Hints to navigators and discoverers on the nomenclature of hydrography and geography. *Oriental Herald*, 20 (1829), pp. 114–123.

—— The alphabets. *Separately paged sections of* The Bible of Every Land. Samuel Bagster and Sons. *London*, [1860].

AUROUSSEAU, M. New Maltese orthography. *Geogr. J.*, 91 (1938), pp. 468–469.

—— Flemish names in Belgium. *Ib.*, 92 (1938), pp. 539–540; transl. in *B. Soc. R. Géogr., Anvers*, 59 (1939), pp. 59–63.

—— A national language for the Philippines. *Geogr. J.*, 92 (1938), pp. 565–566; Italian summary in *B. R. Soc. Geogr. Italiana*, Ser. vii, 4 (1938), p. 332; Spanish transl. in *Riv. Geogr. Americana*, vol. 11, Mar., 1939, p. viii.

—— Geographical names in South Africa. *Geogr. J.*, 94 (1939), pp. 45–49.

—— Official romanization of Thai (Siamese). *Ib.*, 98 (1941), pp. 154–155.

—— Transcription of foreign names. *PCGN Glossaries:* 1. *Modern Greek*, 1942, pp. vii–xiii.

—— Transliteration of modern Greek. *Ib.*, pp. xxv–xxix.

—— Transliteration of Russian. *PCGN Glossaries:* 2. *Russian*, 1942, pp. 16–21.

—— Suggested principles for the use and spelling of geographical names. *Geogr. J.*, 100 (1942), pp. 177–181 and 245–256.

—— The writing of geographical names. *PCGN Glossaries:* 3. *Albanian*, 1943, pp. 7–10.

—— On lists of words and lists of names. *Geogr. J.*, 105 (1945), pp. 61–67; transl. in *Estudios geográf.*, 6 (1945), pp. 813–824; and in *Rev. Brasileira de Geogr.*, 8 (1946), pp. 375–378.

—— Unresolved problems in geographical nomenclature. *Scottish Geogr. Mag.*, 63 (1947), pp. 1–10.

—— The name Kaieteur. *Geogr. J.*, 109 (1947), p. 285; and [*in extenso*] in *The Daily Argosy* (Georgetown), Jan. 20, 1947.

—— Names given by Cook. *Geogr. J.*, 111 (1948), p. 143.

—— The treatment of Antarctic names. *Geogr. Rev.*, 38 (1948), pp. 487–490.

—— [Remarks on standardization of geographical names]. *Proc. and Trans. Third Internat. Congr. Topon. and Anthropon., Brussels*, 1949, II, Louvain, 1951, p. 399.

—— What is an English geographical name? *Geogr. J.*, 117 (1952), pp. 191–197.

—— Transliteration of Cyrillic script. *Nature*, 171 (1953), p. 940.

—— Geographical names for international use. (Paper discussed at the Fifth Internat. Congr. Onomastic Sci., Salamanca, 1955. *In the press*).

—— *See* German National Socialist Republic; *see* United Kingdom.

AUSTRALIAN COLONIES: Inter-Provincial Geographical Conference, Melbourne, 1884. Reported by E. J. Forbes, *untitled serial publn., R. Geogr. S. Australasia (Victorian Br.)*, 17, pt. 2, 1900, p. 10.

AUSTRO-HUNGARIAN EMPIRE: Kriegs Ministerium. 'Die geographischen Namen in den Militär-Karten'. C. Ritter von Steeb. *Mitt. k. k. Mil. Geogr. Inst.*, 17, 1897 (1898), pp. 53–66.

—— Die Schreibung der geographischen Namen auf der Balkan-Halbinsel. J. Levačić. *Ib.*, pp. 67–74.

—— Die Schreibung geographischer Namen nach russischen Kartenwerken. F. Bielawski. *Ib.*, pp. 75–79.

BELL, A. M. English visible speech. *Brit. colon. Print.*, 1863.

BERRY, J. The recording of African place names. *Proc. and Trans., Third Internat. Congr. Topon. and Anthropon., Brussels*, 1949, III (Louvain, 1951), pp. 471–477.

—— Recording geographical names in Africa south of the Sahara. (Read at Fifth Internat. Congr. Onomastic Sci., Salamanca, 1955. *In the press.*)

BIELAWSKI, F. *See* Austro-Hungarian Empire.

BONACKER, W. Karten-Wörterbuch. *Berlin*, 1941. (But see *Geogr. J.*, 105 (1945), pp. 64–65.)

BRITISH ACADEMY. Transliteration of Arabic and Persian. *Proc. Brit. Acad.*, 8 (1917–1918), pp. 505–521.

—— Transliteration of Slavonic. *Ib.*, pp. 523–542.

BRITISH AND FOREIGN BIBLE SOCIETY. Literation and transliteration. H. Morris. *Bible House Papers*, 5 (1901), pp. 1–30.

BRITISH ASSOCIATION FOR THE ADVANCEMENT OF SCIENCE: Committee appointed to inquire into the practicability of an international auxiliary language. [Rept.] *In Rept. Brit. Assn. Adv. Sci., Edinburgh*, 1921 (1922), pp. 390–407; [Report to Council, Sept., 1942], *Nature*, 150 (1942), pp. 718–719.

BRITISH LEGATION, Athens. Memorandum. 'Transliteration of modern Greek'. *Athens*, 1919.

BRITISH MUSEUM. Rules for compiling the catalogues in the Department of Printed Books in the British Museum. *London*, 1920.

BRITISH SCHOOL OF ARCHÆOLOGY, Athens. Notice to contributors. *Ann., Brit. Sch. Archæol.*, Athens, 16 (1909–1910), pp. 332–333.

BURGESS, J. Geographical place-names in Europe and the East. *Rept., Sixth Internat. Geogr. Congr., London*, 1895 (1896), pp. 493–503.

BURRILL, M. F. Official standardisation of geographic names in the United States. *Proc. and Trans., Third Internat. Congr. Topon. and Anthropon., Brussels*, 1949, II (Louvain, 1951), pp. 394–399.

—— Some considerations in international standardisation of geographic names. (Read at Fifth Internat. Congr. Onomastic Sci., Salamanca, 1955. *In the press.*)

CAMBRIDGE UNIVERSITY LIBRARY. Rules for the catalogues of printed books, maps & music. *Cambridge*, 1927.

CHINESE PEOPLE'S REPUBLIC. Chinese language in Latin Script. Literary revolution ready. *The Times*, 3rd May, 1956.

CHINESE REPUBLIC: Directorate General of Posts. 'List of Post Offices'. 13th issue. Shanghai, 1932.

CLARENDON PRESS, *See* Admiralty.

D'ABBADIE, A. L'orthographe des noms géographiques. *C. R., Soc. de Géogr.*, 1884, p. 343.

DALTON, J. C. The orthography of geographical names. *Geogr. J.*, 1 (1893), pp. 431–437.

DAMIANI, E. Sull'unificazione della trascrizione dei nomi slavi, originalmente scritti in caratteri cirillici nei cataloghi delle biblioteche a caratteri latini. *L'Europa Orientale*, 15 (1935), pp. 449–452.

—— Sur l'état actuel des systèmes de transcription des noms slaves cyrilliques dans la documentation bibliographique. *Trans., Internat. Fed. Documentation, 14th Conf., London*, 1938, II, c (1938), pp. 245–248.

DAWKIN, R. M. The transliteration of modern Greek. *Ann., Brit. Sch. Archæology, Athens*, 15 (1908–1909), pp. 214–222.

De-Toni, J. B. Repertorium geographico-polyglottum. *Patavii*, MDCCCXCIV.

Dominion of Canada: Geographic Board. Reports, 1899–1924. (Some contain important special articles.)

——: Board on Geographical Names. Gazetteer of Canada. *Ottawa*, 1952–(*in progress*).

Dominion of New Zealand: Honorary Geographic Board. Bull. 1. Rules of nomenclature and List of Names approved, or changed, or expunged. *Wellington*, 1934.

——: List of Names approved . . . 1924–1946. *Wellington*, 1948.

Ekwall, E. English River Names. *Oxford*, 1928.

—— The Concise Oxford Dictionary of English Place-names. *Oxford*, 1936.

Empire Survey Review. [Correspondence on the spelling of African place-names]. 10 (1949–1950), pp. 92–94, 189–190, 237–239, 283–286.

Encyclopædia Britannica, 11th ed. Slavs. [Table of Slavonic alphabets, p. 232).

English Place-Name Society. [Publications]. *Cambridge*, 1924–(*in progress*). See especially vol. I, pt. 2, The Chief Elements used in English Place-names.

French Republic: [Conventional alphabet]. *B. Soc. Géogr.*, Sér. vii, 7 (1886), pp. 192–202.

——: Troupes d'Occupation du Maroc. 'Règles de transcription des noms indigènes suivies de vocabulaires arabe et berbère'. *Casablanca*, 1917.

General Register Office. Census 1951, England and Wales. 'Index of Place-names'. 2 vols., *London*, 1955.

German Empire: [Conventional alphabet]. *Ann. der Hydrogr.*, 1888, pp. 543–544.

——: Auswärtiges Amt. 'Grundsätze für die Namengebung, Namenübersetzung, Schreib- und Sprechweise der geographischen Namen in den deutschen Schutzgebieten'. *Deutsches Kolonialbl.*, 14 (1903), p. 453.

GERMAN NATIONAL SOCIALIST REPUBLIC. Richtlinien für die Darstellung der Reichsgrenzen vom Jahre 1918 und der Orts- und geographischen Namen in den abgetrennten und Auslandsgebieten. *Mitt. Reichsamt f. Landesaufn.*, 14 (1938), pp. 123–129; transl. by M. Aurousseau in *PCGN Circ. No. 5*, 30th Aug., 1938 (unpublished).

—— Wörterverzeichnis für russische Karten. H. P. Kosack. *Mitt., Reichsamt f. Landesaufn., Sonderheft* 23, 1943.

GILCHRIST, J. B. A Dictionary, English and Hindoostanee. *Calcutta*, 1787–1790.

GLEICHEN, Lord EDWARD. The decisions of the Permanent Committee on Geographical Names on the transliteration of Arabic characters. *Geogr. J.*, 56 (1920), pp. 308–313.

—— The Permanent Committee on Geographical Names. *Ib.*, 57 (1921), pp. 36–43.

—— The spelling and pronunciation of geographical names for British official use. *Rept., Internat. Geogr. Congr., Cambridge*, 1928 (1930), pp. 445–451.

—— Some recent decisions of the P.C.G.N., *Geogr. J.*, 77 (1931), pp. 161–163.

—— *See* United Kingdom.

—— and J. H. Reynolds. Permanent Committee on Geographical Names for British Official Use. *Geogr. Teacher*, 12 (1923–1924), pp. 448–449.

GRAESSE, J. G. Th. Orbis latinus. 2nd ed., *Berlin*, 1909.

HALHED, N. B. A Code of Gentoo Laws. *London*, 1776.

HERSCHEL, Sir J. *See* Admiralty.

HINKS, A. R. [On geographical names, and on glossaries of pronunciation]. *Geogr. J.*, 46 (1915), pp. 31–37 and 48–50.

—— Names in Manchuria. *Ib.*, 52 (1918), pp. 311–314.

—— Mountain names on the Indian border. *Ib.*, 74 (1929), pp. 274–277.

—— Relative terms for geographical position. *Ib.*, 79 (1932), pp. 541.

—— The P.C.G.N., *in* Hints to Travellers, 11th ed., vol. 2, *London*, 1938, 214–218.

—— The prime meridian. *Geogr. J.*, 101 (1943), pp. 47.

—— [On physical names for the map of Britain]. *Ib.*, 102 (1943), pp. 159–160.

H.M. POSTMASTER GENERAL. Post Offices in the United Kingdom and the Irish Republic [*London*], Oct., 1955.

HUNTER, Sir W. W. Letter to the Secretary of the Government of India, Home Department. *Gazette of India*, 31st May, 1870.

—— Guide to the orthography of Indian proper names, with a list showing the true spelling for all post towns and villages in India. *Calcutta*, 1871.

HUTT, J. B. Foreign geographic names on the mariner's chart. *U.S. Naval Inst. Proc.*, 72 (1946), pp. 39–47.

INDIAN EMPIRE: Dept. of Revenue, Agriculture and Commerce. [Resolution of 20th Aug., 1873, on official adoption of the Hunterian System], *in* A Manual of Surveying for India, by H. L. Thuillier and R. Smith, 3rd ed., *Calcutta*, 1875, pp. 408–411.

——: General Staff. A system for the transliteration of Persian, Afghan and Arabian words. 2nd ed., reprinted, *Calcutta*, 1916.

——: Survey of India. Rules for transliteration, *in* Handbook of Professional Instructions for the Topographical Branch, by St. G. C. Gore and C. Strahan, 2nd ed., 1896, para. 78, pp. 42–44. [States that the system adopted by the Govt. of India was 'a modification of that proposed by Dr. Hunter'.]

——: ——. Glossary of words in various Indian and other Asiatic languages used on Survey of India maps. *Calcutta*, 1940.

——: ——; *cf.*, Wauhope.

INTERNATIONAL CIVIL AVIATION ORGANISATION. Abbreviations of place-names. *Doc.* 6919–*Com*/532/2, *Amendment No. 3*, [Montreal], 31st Dec., 1954.

INTERNATIONAL CONGRESS OF ORIENTALISTS. Rapport de la Commission de Transcription. Suppt. to *Actes, Dixième Congr. Internat. Orient., Genève*, 1894 (Leide, 1897); transl. in *J. R. Asiatic Soc.*, 1895, pp. 879–892.

INTERNATIONAL FEDERATION FOR DOCUMENTATION. Transliteration and transcription. *Trans., 14th Conf., London*, 1938, II, c (1938), pp. 244–245.

[INTERNATIONAL GEOGRAPHICAL CONGRESS]. *C. R., Congr. des Sci. géogr., cosmogr. et comm., Anvers*, 1871, 2 (1872), p. 160.

—— *Notiz. e Rend., Terzo Congr. Geogr. Internaz., Venezia*, 1881, 1 (1882), p. 326.

—— *C. R., Vme Congr. Internat. Sci. Géogr., Berne*, 1891 (1892), pp. 112, and 492 (footnote).

—— *Rept., Sixth Internat. Geogr. Congr., London*, 1895 (1896), pp. 511–512.

—— Rules for geographical nomenclature. *Verh., siebenter Internat. Geogr. Kongr., Berlin*, 1899, 1 (1901), p. 314.

—— *C. R., Neuvième Congr. Internat. Geogr., Genève*, 1908, 1 (1909), pp. 149–150.

—— Rapporto della commissione per la trascrizione dei nomi geografici. G. Ricchieri. *Atti, X Congr. Internaz. Geogr., Roma*, 1913 (1915), p. 100.

INTERNATIONAL GEOGRAPHICAL UNION. *C. R., Congr. internat de géogr., Amsterdam*, 1938, 1 (1939), p. 500.

INTERNATIONAL HYDROGRAPHIC BUREAU. Glossary of cartographic terms and manual of symbols and abbreviations. *Spec. Publn. No. 22*, 3rd ed., Monaco, 1951.

INTERNATIONAL HYDROGRAPHIC CONFERENCES, 1919, 1926 and 1929. Resolution No. I, 27, *in* Repertory of the technical resolutions. *Monaco*, 1930.

INTERNATIONAL LIBRARY COMMISSION. 8th Meeting, Madrid–Barcelona, 1935. *Actes, Publ., Fédération Internat. des Bibliothécaires*, 7 (1935).

INTERNATIONAL MAP COMMITTEE. Resol. and Proc., Internat. Map Committee assembled in London, Nov., 1909. *London*, 1910, p. 4.

—— C. R., deuxième conf. internat., Paris, 1913. *Paris*, 1914, pp. 87–89.

INTERNATIONAL PHONETIC ASSOCIATION. The Principles of the International Phonetic Association. *Maître phonétique*, Sept.–Oct., 1912, suppt.

INTERNATIONAL ORGANISATION FOR STANDARDISATION. Recommendation R 9. 'International System for the transliteration of Cyrillic characters'. 1st ed., Oct., 1955.

JAPANESE EMPIRE: Hydrographic Department, I.J.N. 'Tables of Japanese Place Names'. *N. P.*, 1936, (In Japanese); reproduced as Gazetteer of the Japanese Empire, *U.S. Hydrographic Office, Misc. No. 10947*, Aug., 1943.

JONES, D., and A. Camilli. Fondamendi di grafia fonetica. 20 pp., *London*, 1933.

[JONES, Sir W.] A dissertation on the orthography of Asiatick words in Roman letters. *Asiatick Researches* (*Tr. Asiatick Soc.*) 1 (1788), pp. 1–56.

KINGDOM OF BELGIUM. [Conventional alphabet]. *Mouvement géogr.*, 1892, p. 25.

KINGDOM OF ITALY: Ministero delle Colonie. 'Norme per la trascrizione italiana e la grafia araba dei nomi propri geografici della Tripolitania e della Cirenaica'. *Roma*, 1915.

KINGDOM OF SPAIN. [Conventional alphabet]. *B. Soc. Geogr. Madrid*, 1 (1876), pp. 88–98.

KNOX, A. *See* War Office.

KÖPPEN, W. Die Schreibung geographischer Namen. Vorschlag an den Deutschen Geographentag von 1893. *Hamburg*, 1893. (Original not seen. Reviewed by J. C. Dalton, *q.v.*)

—— Über Schreibung russischer geographischen Namen. *Mitt. Reichsamt f. Landesaufn.*, 15 (1939), pp. 151–157.

LANGENFELT, G. On the origin of tribal names. *Anthropos*, 14–15 (1919–1920), pp. 295–313.

—— Geografisk nationalism och internationell handel. *Index*, 8 (1933), pp. 142–152.

LATIFI, A. A national script for India. *Asiatic Rev.*, July, 1929.

LEVAČIĆ, J. *See* Austro-Hungarian Empire.

LEWIN, E. Publications in the Library of the Royal Empire Society on place names in the overseas Empire. *B. Inst. Hist. Research*, 16, No. 48, Feb., 1939, pp. 168–180.

LIBRARY ASSOCIATION. Cataloguing Rules. *London*, 1930.

劉鈞仁 (Liu Chün-Jen) 中國地名大辭典 (Great Dictionary of Chinese Place-Names). *Peiping*, 19th Year of the Republic.

Маркс, А. Ф. (Marks, A. F.) *Index to* Болшой всемірный настольный атлас маркса (Marks's Great Universal Reference Atlas). *St. Petersburg*, 1905.

Mason, K. Karakoram nomenclature. *Geogr. J.*, 91 (1938), pp. 123–128; and *Himalayan J.*, 10 (1938), pp. 86–90.

Monier-Williams, M. Original papers illustrating the history of the application of the Roman alphabet to the languages of India. Ed. by M. Monier-Williams. *London*, 1859.

—— The duty of English-speaking Orientalists in regard to united action in adhering generally to Sir William Jones's Principles of Transliteration. *J. R. Asiatic Soc.*, 1890, pp. 607–638.

Morris, H. *See* British and Foreign Bible Society; *see* Royal Asiatic Society.

Müller, F. Max. Proposals for a missionary alphabet. *London*, 1854.

Mutziger, J. Roman-letter spellings of toponyms from other writing systems. *Proc. and Trans., Third Internat. Congr. Topon. and Anthropon., Brussels*, 1949, II (Louvain, 1951), pp. 400–403.

Myres, J. L. The word 'international'. *Nature*, 164 (1949), p. 282.

Ordnance Survey. Topography in the tropical forest belt. Training Ser., No. 2, *Southampton*, 1931.

—— The national plans. H. St. J. L. Winterbotham. *Prof. Papers, N.S.*, 16 (1934), p. 58.

—— Gazetteer of Great Britain. *Chessington*, 1946, reprinted 1951.

—— Glossary of . . . Gaelic and Scandinavian elements . . . in place names of Scotland. *Chessington*, 1951.

Oxford University Press. Rules for compositors and readers. 31st ed., *London*, 1938.

Panikkar, K. M. India and the Indian Ocean. *London*, 1945.

PARMENTIER, T. Quelques observations sur l'orthographe des noms géographiques. *C.R. Ass. franç. Av. Sci.*, 1877 (1878), pp. 1015–1059.

—— De la transcription pratique, au point de vue français, des noms arabes en caractères latins. *Ib.*, 1879 (1880), pp. 1067–1109.

—— Vocabulaire arabe-français des principaux termes de géographie et des mots qui entrent le plus fréquemment dans la composition des noms de lieu. *Ib.*, 1881 (1882), pp. i–l.

—— Vocabulaire magyar-français des principaux termes de géographie et de topographie, ainsi que des mots, etc. *Ib.*, 1882 (1883), pp. 921–965.

—— Vocabulaire turk-français des principaux termes de géographie et des mots, etc. *Ib.*, 1883 (1884), pp. 859–931.

—— Vocabulaire scandinave-français des principaux termes de géographie et des mots, etc. *Issued separately by the* Assn. franç. Av. Sci., *Paris*, 1887.

—— Vocabulaire rhétoroman des principaux termes de chorographie et des mots, etc. *Issued separately by the* Assn. franç. Av. Sci., *Paris*, 1896.

POLLACCHI, P. Lecture des cartes russes. *Paris*, 1907.

—— Lecture des cartes anglaises et des Etats-Unis. *Paris*, 1908.

—— Alphabets en usage dans les principales langues parlées dans l'Europe centrale et les Balkans. *N.P.*, 1910.

RAILWAY AND COMMERCIAL GAZETTEER of England, Scotland and Wales. 19th ed., *London* [1938].

RAZMARA, H. A. Note à propos de quelques noms géographiques et de l'alphabet des cartes. Téhéran, le 1er Mai, 1950.

REYNOLDS, J. H. Place-names in the Irish Free State. *Geogr. J.*, 63 (1924), pp. 153–155.

—— Place-names in Sinkiang. *Ib.*, 65 (1925), pp. 242–247.

—— The question of pronunciation tables for the British sheets of the International Map. *Ib.*, 67 (1926), pp. 57–63.

—— The official romanization of Japanese. *Ib.*, 72 (1928), pp. 360–362.

—— Nationalities in the U.S.S.R. *Ib.*, 73 (1929), pp. 370–374.

—— The new orthography in Turkey. *Ib.*, 74 (1929), pp. 72–74.

—— *Review of* Transliteration: Arabic-Hebrew-English (Jerusalem, 1931). *Ib.*, 78 (1931), p. 312.

—— The name Hispaniola. *Ib.*, 85 (1935), pp. 107–108.

—— *See* Gleichen; *see* United Kingdom.

RICCHIERI, G. *See* International Geographical Congress.

RITTER VON STEEB, C. *See* Austro-Hungarian Empire.

RIVERS, W. H. R. Island-names in Melanesia. *Geogr. J.*, 39 (1912), pp. 458–468.

ROYAL ASIATIC SOCIETY of Bengal. Instructions to authors for the submission of papers for publication in the Journal and Memoirs of the Society. *Year-book for 1940* (1941), pp. 195–199.

ROYAL ASIATIC SOCIETY of Great Britain and Ireland: Transliteration Committee. Report. (Adopted by Council, 8th May, 1894.) 13 pp. and two folded tables. Issued with *J. R. Asiatic Soc.*, 1894.

—— Transliteration. Issued with *Ib.*, 1896.

—— Alphabet for languages not yet reduced to writing. H. Morris. *Ib.*, 1898, pp. 23–28.

—— Transliteration of the Sanskrit, Arabic, and allied alphabets. Appendix to *Ib.*, yearly, 1900–1921; reprinted in *Ib.*, 1924, and at irregular intervals thereafter.

ROYAL GEOGRAPHICAL SOCIETY. [Footnote on the orthography of geographical names]. *J. R. Geogr. Soc.*, 6 (1836), p. 6.

—— System of orthography for native names of places. [The R.G.S. System]. *Proc. R. Geogr. Soc.*, 7 (1885), pp. 535–536; revised, *Ib.*, 14 (1892), pp. 116–119; two new rules added, *Geogr. J.*, 2 (1893), pp. 456–457; suppt. on Chinese and Russian issued [in 1894]; system revised, *R. Geogr. Soc. Year-book and Record*, 1900, pp. 35–40; revised and issued as leaflet, 1905; reprinted, 1917.

—— On the giving of names to newly-discovered places. *R. Geogr. Soc. Year-book and Record*, 1901, p. 42; *and in* Hints to Travellers, 8th ed., vol. 2, *London*, 1901, p. 256a; reprinted in 9th and 10th eds.

—— The Society's rules for the orthography of native names. *Circular Memorandum*, 11th May, 1914.

—— New geographical names, *in* Hints to Travellers, 11th ed., vol. 2, *London*, 1938, pp. 209–214.

ROYAL SCOTTISH GEOGRAPHICAL SOCIETY: [Committee on the place-names of Scotland]. *Scottish Geogr. Mag.*, 7 (1891), p. 688; 29 (1913), pp. 54, 268–269, 322–323; 54 (1938), pp. 148–150, 383; 56 (1940), p. 144; 62 (1946), p. 144; 69 (1953), p. 144.

——: Bibliography of the place-names of Scotland. *Edinburgh*, 1945.

ROYAL SOCIETY. The transliteration of Russian, Serbian and Bulgarian for bibliographical purposes. *London*, 1953; (but see *Nature*, 171 (1953), p. 940).

SCHMIDT, P. W. Die Sprachlaute und ihre Darstellung in einem allgemeinen linguistischen Alphabet. *Anthropos*, 2 (1907), pp. 282–329, 508–587, 822–897, 1058–1105.

SCHOMBURGK, Sir R. On a uniform system to reduce unwritten languages to alphabetical writting in Roman characters. *Rept., Brit. Assn. Adv. Sci., Swansea*, 1848, (1849), p. 99.

SCHOOL OF RUSSIAN STUDIES (Liverpool). [Scheme of two alphabets for transliteration into English]. *In* The Russian Year-book for 1912, *London*, 1912, p. 768.

SCHOOL OF SLAVONIC AND EAST EUROPEAN STUDIES (London). Conference [Oct. 1921] of University teachers of Russian and other Slavonic languages. *Slavonic Rev.*, 1 (1922), pp. vi–viii.

SOCIETA GEOGRAFICĂ ROMÎNĂ. Marele Dicționar Geografic al Romîniei. 5 vols., *București*, 1898–1902.

SOCIÉTÉ ASIATIQUE. Transcription des alphabets orientaux. Suppt. to *J. Asiatique*, ccii, Jan.–Mars., 1923, 16 pp.

SOCIETY FOR THE PROMOTION OF HELLENIC STUDIES. Notice to contributors. *J. Hellenic Stud.*, 23 (1903), pp. lv–lvi.

STATESMAN'S YEAR-BOOK, The. *London*, yearly since 1863.

丁文江 (Ting, V. K.) 中華民國新地圖 (New Atlas of the Chinese Republic), *Shanghai* 1934; *index reproduced as* Gazetteer of Chinese place names based on the index to V. K. Ting Atlas. Compiled by the U.S. Board on Geographical Names. *Washington, D.C.*, June, 1944.

TUCKER, A. N. The spelling of African place-names on maps. *B. Sch. Orient. Afr. Stud.* (London), 12 (1948), pp. 824–830.

—— The spelling of African place names. *Empire Surv. Rev.*, 10 (1949–1950), pp. 332–335.

UNION OF SOUTH AFRICA: Departmental Committee on the form and spelling of geographical proper names. Report. *Pretoria*, 1939.

——: Place Names Committee. 'Official Place Names in the Union and South West Africa'. *Pretoria*, 1951.

UNION POSTALE UNIVERSELLE. Dictionnaire des Bureaux de Poste. 5th ed., 2 vols., *Berne*, 1951.

UNITED KINGDOM: PERMANENT COMMITTEE ON GEOGRAPHICAL NAMES. Rules for the spelling of geographical names for British official use. (Termed the R.G.S. II System). *Geogr. J.*, 57 (1921), pp. 39–43; reissued as a leaflet, Jan., 1921; 2nd ed., July, 1921; 3rd ed., 1924; 4th ed. (revised), 1932; 5th ed. (corrected), *Geogr. J.*, 97 (1941), pp. 318–321.

——: —— Alphabets of Foreign Languages. Lord Edward Gleichen and John H. Reynolds. London, 1921; 2nd ed., 1933; reprinted with . . . certain additions by Marcel Aurousseau, 1944; reissued 1951.

——: —— [Lists of names by countries], *London*, 1921–1938.

——: —— Leaflets [on transliteration], 1925–(*in progress*).

——: —— Glossaries, 1942–(*in progress*). Those for Modern Greek, Russian, Albanian, Serbo-Croat, Romanian, Siamese, Turkish and Japanese have appeared.

——: —— Gazetteer of Greece. *London*, 1942.

——: —— Gazetteer of Albania. Provisional ed., London, 1944, complete ed., 1946.

——: —— Principles of Geographical Nomenclature. 4 pp., *London*, 1954.

——: —— Lists of names (new series). London, 1954–(*in progress*).

——: —— Organisation and functions. *Geogr. J.*, 122 (1956), pp. 393–394.

——: —— *cf.*, Admiralty, British Museum, General Register

Office, H.M. Postmaster General, Ordnance Survey, Railway, War Office.

——: —— *See also* Aurousseau, Gleichen, Hinks, Reynolds.

UNITED NATIONS. Nomenclature of geographic areas for statistical purposes. *Statist. Papers, Ser. M., No. 1*, Lake Success, N.Y., 1949.

——: Economic and Social Council. Resolutions adopted by the U.N. Regional Cartographic Conference for Asia and the Far East. *E/Conf. 18/5 Rev. 1*, 8th July, 1955.

UNITED STATES: Geographic Board. Reports. *Washington, D.C.*, 1892–1932.

——: First report on foreign geographic names. Washington, 1932.

——: Board on Geographical Names. Decisions, Washington, 1936–1947.

——: —— Guide to geographical names in China. *Spec. Pubn., No. 24*, 1944.

——: —— Guide to geographical names in the Japanese Empire. *Ib.*, No. 25, 1944.

——: —— Guide to geographical names in Korea (Chōsen). *Ib.*, No. 51, 1945.

——: —— Transliteration of Arabic and Persian. *Ib.*, No. 78, 1946.

——: Board on Geographic Names. Decisions, 1947–(*in progress*).

——: —— The geographical names of Antarctica. *Spec. Pubn., No.* 86, 1947; suppt., 1949.

——: —— Gazetteers. 1955–(*in progress*).

——: *See also* Burrill, Hutt, Mutziger, Ting.

——: Foreign Broadcast Information Service. Transliteration Manual. *Washington, D.C.*, Dec., 1951.

——: Government Printing Office. Style Manual. Rev. ed., *Washington, D.C.*, Jan., 1945.

——: Hydrographic Office. No. 98. Report on uniform system of spelling foreign geographical names. *Washington*, 1891.

——: —— *See also* Japanese Empire.

Виноградов, Н.В. (Vinogradov, N. V.). Карты и Атласы (Maps and Atlases). *Moscow and Leningrad*, 1941, p. 44.

WADE, Sir T. F., and W. C. Hillier. Tzŭ Êrh Chi. 2nd. ed., 3 vols., *Shanghai*, 1886.

WAR OFFICE. Hints on reconnaissance mapping for explorers in unsurveyed country. 7 pp., [*London*, 1892].

—— Rules for the transliteration of place-names occurring on foreign maps. A. Knox. T.S.G.S., *London*, 1906; reprinted 1919.

—— List of some Chinese words used in geographical nomenclature. *In* GSGS No. 2165, 1908.

—— Glossary of typographical (*sic*) and descriptive terms in Albanian, Bulgarian, Czeck (*sic*), French, German, Greek, Hungarian, Italian, Polish, Rumanian, Russian, Serbian, Turkish. GSGS No. 2831, *London*, 1917.

—— G.S.G.S. Short Glossaries for use on foreign maps. *London*, 1943–(*in progress*. Covers about forty languages).

WASHINGTON, J. *See* Admiralty.

WAUHOPE, R. A. Notes on the transliteration of place-names from Persian and Arabic into Roman script. *Dehra Dun*, 1918.

—— Notes on the spelling of Turki, Tibetan, and Chinese place-names. *Dehra Dun*, 1919.

WHARTON, L. C. *See* International Federation for Documentation.

WHEELER, G. F. Latinization: a study in Middle Eastern language reform. *J. R. Central Asian Soc.*, 24 (1937), pp. 591–604.

—— [Papers on the learning of languages, by Karshish]. *J. United Serv. Inst., India*, 70 (1940), pp. 87–94, 161–176, 259–271, 362–374; 71 (1941), pp. 35–47 and 129–140; *cf.*, *Geogr. J.*, 100 (1942), p. 274.

WHITAKER'S ALMANACK. *London*, yearly since 1868.

WINTERBOTHAM, H. St. J. L. *See* Ordnance Survey.

ANALYTICAL INDEX